LAW AND SOCIAL NORMS

Law and Social Norms

ERIC A. POSNER

Harvard University Press
Cambridge, Massachusetts
London, England 2000

Printed in the United States of America

Library of Congress Cataloging-in-Publication Data

Posner, Eric A.
Law and social norms / Eric A. Posner.
p. cm.
Includes bibliographical references and index.
ISBN 0-674-00156-7 (hard cover)
1. Sociological jurisprudence. 2. Law and economics. 3. Social norms.
4. Collective behavior. 5. Game theory. I. Title.
K370.P67 2000
340'.115—dc21 99-088226

To my parents ☋

Contents

LAW AND SOCIAL NORMS

1

Introduction: Law and Collective Action

A married woman who is pregnant asks a court to order a blood test of the alleged father (not her husband) for the purpose of establishing his paternity. The court rejects the woman's motion, relying on the common law presumption of the legitimacy of a child whose mother is married. The court says that the presumption reduces the chance that a child will be stigmatized as illegitimate. The implication is that it is better to spare a child the stigma of illegitimacy than allow him to learn the identity of his biological father.[1]

Police warn local merchants that a certain individual has been arrested for shoplifting, though was never prosecuted. A school warns a potential employer that one of its teachers has sexually harassed a student, though this teacher was not disciplined, dismissed, or fired. The transmission of accurate information surely benefits the merchants and the potential employer of the teacher, but it also may stigmatize. Does this possibility trigger the requirement of due process?[2]

A tugboat operator fails to equip its tugboats with working radios, so the tugboats do not receive weather reports that might have enabled them to avoid a storm that damages customers' goods. The tugboat operator argues that because there was no custom in the tugboat industry to use radios, it should not be held negligent. The court holds that custom is no defense.[3] But why not? Are customary business practices likely to be inferior to those dictated by courts?

A wife and a husband obtain a divorce in civil court. The husband and wife's religious congregation recognizes divorces only if the husband gives his consent, and would ostracize the wife if she remarried in the absence of a religiously sanctioned divorce. The wife submits to a one-sided distribution of assets because the husband threatens to withhold his consent to a religious divorce. Later she asks a civil court to void the contract on the ground that she signed it under duress.[4] If the court grants her request, what effect would her victory have on the cohesiveness of the congregation?

A frequently offered justification for *Brown v. Board of Education*[5] is that even if appropriate separate facilities for blacks were supplied by states, a policy of separating blacks and whites is unconstitutional because it stigmatizes blacks.

A court prohibits a local government from constructing a Christmas tree or crèche on public land. The project offends many people, but it does not hurt them, while it pleases many more people. People are offended by countless projects, from sex education to gun control, but offense of this sort is rarely a decisive factor in the evaluation of such projects. What is different about religious symbolism?

Every day courts must evaluate the stigmatizing effect of an action, the conformity of behavior with social norms, the meaning of symbols, and the consequences of ostracism. Is the stigmatizing of AIDS patients, or illegitimate children, or recipients of welfare, a simple injury like a poke in the eye, or does it enhance social norms that contribute to public order? When businesses follow customary practices rather than deviating, should we assume that the customs reflect the march of progress or the stampeding of sheep? When the government engages in symbolic behavior, like the construction of idols, or restricts symbolic behavior, like the desecration of flags, can it be the case that there is nothing substantive at stake? Can legal intervention eliminate stigmas, change customs and social norms, transform the meaning of symbols—or are these social facts unyielding in the face of self-conscious attempts at reform?

That these things do matter in law and politics cannot be denied. Symbolism and stigma play a role in every major piece of legislation. Flag desecration bills are designed to rebut the symbolism of destroying flags. Affirmative action enhances the stigma of belonging to a minority, according to its opponents, or weakens that stigma, according to its defenders. Modern social welfare and bankruptcy legislation was intended to eliminate the stigma against people who are poor and cannot pay their debts, and against aliens and illegitimate children, yet earlier versions of this legislation were intended to strengthen the stigma. Expungement laws, which erase criminal convictions from offenders' records, reduce the stigma of the ex-convict. And debates over issues as diverse as the sale of organs, surrogate motherhood, the legalization of prostitution, cost-benefit analysis, and pornography invariably raise questions about the symbolism of the practice in question and the use of law to control it.

In a world with no law and rudimentary government, order of some sort would exist. So much is clear from anthropological studies. The order would appear as routine compliance with social norms and the collective infliction of sanctions on those who violate them, including stigmatization of the deviant and ostracism of the incorrigible. People would make symbolic commitments to the community in order to avoid suspicions about their loyalty. Also, people would cooperate frequently. They would keep and rely on promises, refrain from injuring their neighbors, contribute effort to public-spirited projects, make gifts to the poor, render assistance to those in danger, and join marches and rallies. But it is also the case that people would sometimes breach promises and cause injury. They would discriminate against people who, through no fault of their own, have become walking symbols of practices that a group rejects. They would have disputes, sometimes violent disputes. Feuds would arise and might never end. The community might split into factions. The order, with all its benefits, would come at a cost. Robust in times of peace, it would reveal its precariousness at moments of crisis.

Now superimpose a powerful and benevolent government with the ability to make and enforce laws. Could the government selectively intervene among the continuing nonlegal forms of order, choosing to transform those that were undesirable while maintaining those that were good? Could it tinker with the incentives along the edges, using taxes, subsidies, and sanctions to eliminate, say, the feuds and the acts of discrimination, without interfering with neighborly kindliness and trust? Or would the sheer complexity of social organization overwhelm such efforts?

Turning from positive to normative, how should legislatures and courts deal with phenomena like stigma and ostracism, social norms, reputation, symbolism, and the numerous other sources of order that exist outside the laws of the state? Should we assume that these phenomena are desirable and should be respected or enhanced, or that they are pathological, and should be deterred? Can we specify conditions under which the state should intervene? Can we evaluate different kinds of intervention according to the likelihood that they will enhance desirable forms of nonlegal cooperation and subvert undesirable forms?

These questions are old, and they have occupied the best minds working in many academic disciplines. But they are largely ignored by mainstream legal scholars writing about how law affects behavior, and even more so by scholars writing about the appropriate direction for legal reform.

This book is about the relationship between law and what I will inelegantly call "nonlegal mechanisms of cooperation." It is motivated by a lacuna in legal theory, and particularly law and economics, on which I draw. The positive branch of law and economics assumes that the individual goes about satisfying his preferences, subject to a budget constraint, but unaffected by the attitudes of others. Preferences may be egoistic or altruistic or both, but nothing, other than the state, prevents individuals from preying on each other when it serves their interests. A person will steal, or drive carelessly, or murder, or lie, unless the state erects a deterrent in the form of laws against theft, negligence, murder, and fraud. This description of the world is partly true, but mostly false. Most people refrain most of the time from antisocial behavior even when the law is absent or has no force. They conform to social norms. The question left unanswered by law and economics is why people conform to social norms. Without an answer to that question, one cannot understand the effect of laws on people's behavior.

The normative branch of law and economics and most other mainstream normative legal theories treat the government as an exogenous force that intervenes to deter socially costly but privately beneficial behavior or, put differently, to solve collective action problems that arise among citizens. Environmental law, for example, is explained and justified as a deterrent to the private incentive to pollute. Clean air, water, and soil are the collective goods that would result from wise environmental laws. Bankruptcy law preserves the value of assets against the uncoordinated efforts by creditors to enforce their claims. The law of intellectual property enables inventors and authors to recoup the cost of their investments when otherwise imitators would deplete the value of original work. Contract law provides security against broken promises, and tort law protects people from interference with their use of property. But these explanations, while useful and interesting, are not complete. The law is always imposed against a background stream of nonlegal regulation—enforced by gossip, disapproval, ostracism, and violence—which itself produces important collective goods. The system of nonlegal cooperation is always in some ways superior and in other ways inferior to the legal solution, and legal intervention will undermine or enhance the background norms of nonlegal cooperation in complex ways. The desirability of a proposed legal rule, then, does not depend only on the existence of a collective action problem on the one hand, and competently operated legal institutions on the other hand. It also depends on the way nonlegal systems always already address that collective action problem and the extent to which legal intervention would interfere with those nonlegal systems.

In making these claims, this book follows a tradition of work that criticizes legal scholarship for focusing too much on the state, for simplifying the relationship between citizens and the government, and for analyzing simple problems to the exclusion of important and interesting ones. Ellickson (1991) is the most recent and influential representative of this tradition, but the discontent that motivated his work can also be found as far back as the 1960s (Macaulay 1963), and even farther back, in the writings of the legal realists (Llewellyn 1931). The influence of this tradition, however, has been limited by an important failure. This is the failure of the critics to supply a useful analytic framework as an alternative to the methodologies they criticize. Partly because of this failure, the influence of this tradition, when not muted, has not always been positive. Influenced in a vague way by the critics, scholars now use the concept of the "social norm" in a profligate and inconsistent way. Scholars need a methodology that enables a systematic analysis of the relationship between the law and nonlegal mechanisms of cooperation.

This book proposes such a methodology. Part One of the book develops a general model of nonlegal cooperation. The model, which is described in Chapter 2, is a signaling game in which people engage in behavioral regularities in order to show that they are desirable partners in cooperative endeavors. Defection in cooperative endeavors is deterred by fear of reputational injury, but the signaling behavior independently gives rise to forms of collective action that can be of great significance. People who care about future payoffs not only resist the temptation to cheat in a relationship; they signal their ability to resist the temptation to cheat by conforming to styles of dress, speech, conduct, and discrimination. The resulting behavioral regularities, which I describe as "social norms," can vastly enhance or diminish social welfare. The analysis is intended to explain crucial concepts that are neglected or misused in law and economics, including the concepts of trust, status, group solidarity, community, social norm, and custom, and to evaluate the relationship between these phenomena and the law.

The model is based on work in game theory and economics that spans the last four decades but has only recently entered the mainstream of economics and still has not had much impact on legal theory. Much of this work addresses two problems: how people are able to cooperate in the face of the incentives to free-ride on group endeavors; and why behavior has a sticky, discontinuous, or norm-driven quality when, under standard economic premises, behavior should reflect people's idiosyncratic and (by assumption) continuously distributed preferences. The most useful approaches to these questions have made strong assumptions about the content of people's utility functions, for example, assuming altruism, envy, or a desire to conform (for example, Frank 1988, Akerlof 1984, ch. 8, S. Jones 1984, Bernheim 1994); appealed to historical background and institutional detail (North 1990); relaxed the usual rationality assumptions, relying instead on the importance of learning and imitation (Young 1998b); and placed weight on the effects of information asymmetries (Spence 1974). In Chapter 3, I explain why I find the last approach to be most useful.

Part Two applies the model to several areas of the law. It begins, in Chapter 4, with the claim that the gift is a fundamental signal. Friends, family members, merchants, politicians, diplomats—all those seeking or participating in reciprocal relationships—give gifts in order to attract new partners and reassure old partners of their continuing commitment. This view contrasts with the conventional view that gift-giving is motivated by altruism, and it sheds light on many puzzles in the legal treatment of gifts and gratuitous promises, such as why gratuitous promises are given less protection than commercial promises.

Another important signal is the wedding vow, but this is only the most conspicuous of the many signals that pass between people who enter and maintain intimate relationships, and between them and their friends, their families, and members of the public. Other signals include the shunning of people who enter non-standard intimate relations, and their children, who used to suffer from the stigma of illegitimacy. Chapter 5 uses the signaling model to discuss family relationships and family law. The signaling model sheds light on such important features as the mandatory structure of marital obligations and the state's reluctance to enforce or interfere with intra-marital agreements.

The wedding vow is an example of a signal that starts off as a form of private behavior but is subsequently institutionalized and regulated by the state. Because signaling is important behavior, the state has powerful incentives to exploit it when it produces benefits, and suppress it when it does not. Another arena in which these incentives are displayed is that of criminal punishment, the subject of Chapter 6. This chapter argues, among other things, that punishments designed to shame criminals are unlikely to produce optimal deterrence, and that they have perverse effects in countries like the United States, where criminal punishments can become badges of status in communities that do not trust the government. The signaling model also reconciles two opposing approaches in criminology, one of which holds that criminal punishments are best understood as prices imposed on criminal behavior, and the other of which holds that people obey the law when they believe that the law is legitimate.

Chapter 7 turns to politics, arguing that certain actions, including self-censorship, respect for the flag, and voting, are ways in which people signal their loyalty to the government or to dominant political groups. This chapter shows how beliefs about the symbolic value of behaviors arise endogenously from a model in which people seek to cooperate for personal gain. This chapter proposes a solution to the "voting paradox"—the paradox that people vote even though any non-tautologous description of the gains from voting indicates that these gains are less than the costs. Chapter 8 extends this analysis to racial discrimination and nationalism. Racism and nationalism are best modeled not as matters of "taste," as under the standard economic approach, but as attitudes that emerge endogenously in a game in which people signal their loyalty to each other by shunning outsiders. The chapter briefly discusses the merits and disadvantages of affirmative action and anti-discrimination law.

Chapter 9 advances a hypothesis about contract law. This chapter argues that nonlegal mechanisms of cooperation support relationships, both business and personal, which legal intervention can undermine. Contract law does not enhance cooperation by punishing those who break their promises, but supplies a commitment mechanism that allows parties to reduce the payoff from breach by giving each other the power to inflict mutual losses at any time and for any reason. Contract doctrines are best understood as a means of enabling parties to make this commitment only when it serves their interest, and of minimizing the size (within constraints) and the variance of the mutually inflicted losses. The theory justifies the highly formal system of contract law that has long been reviled by legal academics.

Part Three turns from particular areas of the law to general issues of normative legal theory. Chapter 10 discusses law and social norms from the perspective of efficiency, redistribution of wealth, and autonomy. It criticizes the view that social norms are efficient, and argues that social norms are often dysfunctional. But this chapter also criticizes the increasingly influential argument that the government should self-consciously try to change social norms, arguing that powerful social norms constrain government actors and that efforts to change social norms can produce unpredictable norm cascades. Chapter 11 argues that people often engage in "principled" behavior for strategic reasons, and examines the implications of this claim for public policy. Chapter 12 criticizes the view that there has been a decline of community in the United States, and that law or the market interferes with communities.

This book has three goals. One goal is to show the value of concepts from game theory for understanding legal issues. This is not a textbook or a survey, however (compare Baird, Gertner, and Picker 1994); the argument should be taken as illustrative of the usefulness of game theory. A more ambitious goal is to persuade the reader about the usefulness of the game theory model I have constructed for illuminating a variety of legal issues. The third goal is to persuade the reader of several substantive claims about the relationship between legal and nonlegal forms of regulation. These claims are scattered throughout the book, but several themes emerge, and they are worth mentioning here.

The first theme is that social norms are usefully understood as mere behavioral regularities with little independent explanatory power and little exogenous power to influence behavior. They are the labels that we attach to the behavioral regularities that emerge and persist in the absence of organized, conscious direction by individuals. These behavioral regularities result from the interactions of individuals acting in their rational self-interest, broadly understood (to include altruism and other forms of interdependent utility), a self-interest that drives people to cooperate across all areas of life. The claim that a social norm caused X or Y is an empty claim. The appropriate claim is "individuals seeking a or b interacted in such a way as to produce behavioral regularities X or Y, regularities that we call 'social norms.'" What distinguishes social norms from other behavioral regularities is that departure from them provokes sanctions, but again the sanctions emerge endogenously as a consequence of people acting in their rational self-interest.

The second theme is that many legal rules are best understood as efforts to harness the independent regulatory power of social norms. These efforts sometimes succeed and sometimes fail; what is important to understand is that social norms are unlikely to change as a result of simple, discrete, low-cost interventions by the government, although proposals along these lines are sometimes found in the literature, and that attempts to intervene are risky, because social norms are complex, poorly understood, and sensitive to factors that are difficult to control. Although social norms are constantly changing as a result of decentralized, undirected interactions, the only way for individuals to self-consciously change them in a direction they seek is to violate them. Not just to violate them, but to violate them in a public and decisive way. Many people engage in this highly risky norm entrepreneurship, but government officials, who do not stand outside the social world, are in a particularly vulnerable position. They are far more likely to conform to social norms than to violate them, so the government will rarely manage radical change of social norms.

The third theme is that many social norms contribute to social welfare, many social norms harm social welfare, and the value of norms is mostly a matter of historical accident. The fourth theme is related to a historical claim that will be familiar to students of sociology, namely, that there has been a gradual displacement of nonlegal regulation by legal regulation. The theme is that this displacement occurred in part because legislatures and courts sought to eliminate the pathologies produced by social norms, not because the social norms reflected values or interests that they do not share. The fifth theme is that this displacement should not be deplored, as is currently fashionable, but celebrated.

PART ONE

Models of Nonlegal Collective Action

2

A Model of Cooperation and the Production of Social Norms

Suppose that a firm seeks an employee for a one-year term, and anticipates that the employee will receive training for the first six months and use his skills to generate profits for the firm during the last six months. During the period of training, the firm will suffer a net loss that it hopes to recover during the period of work. The firm's problem is that if it pays the worker a monthly wage, the worker might quit just when the training is completed, and take himself and his skills to another firm that is willing to pay the higher wage that can be demanded by a skilled worker. The first firm would then either have to hire another, already trained, worker at that higher wage, or pay the existing worker a premium not to leave, in either case losing the return on its investment. Anticipating this problem, the firm does not hire the worker.

Alternatively, the firm might offer to hire the worker on the condition that he receive his entire compensation at the end of the term. Under this arrangement, the worker would not quit after receiving training but before doing work, because then the firm would refuse to compensate him. But the worker might not agree to this arrangement, because he has daily expenses that must be paid from his salary and, moreover, he does not trust the firm to keep its end of the bargain. Instead, the worker might simply promise not to quit prior to the end of his term. But the firm has no reason to believe that the worker will keep his promise. To be sure, many people are honest and will keep their word, but many people are dishonest or adept at rationalizing their actions. Because the firm does not know whether the worker is honest or not, it declines to hire him. A third alternative is for the worker to sign a contract in which he promises not to quit. But the firm knows that courts usually refuse to force people to work by threatening them with jail or other punishments and that, although the firm might be entitled to damages from the worker if the worker quits, it is likely that the worker will be judgment-proof and that the costs of legal proceedings will exceed the damages. The offer to sign a contract therefore would not persuade the firm to hire the worker.

The worker might make the following argument. If he quits at the end of six months, he will do so only in order to obtain a better paying job from another firm. The second firm, however, is likely to know that the worker breached his contract with the first firm. The second firm thus may believe that the worker is unreliable, a person who does not keep his promises. Because the second firm does not want to entrust its valuable assets to a worker who is unreliable, it would not employ such a worker. But if the worker cannot obtain a higher-paying job from a second firm, then he will not breach his contract with the first firm. If the first firm is persuaded by this logic, which depends on, among other things, a network of communication among firms, it will hire the worker—and they may even forgo the formalities of a contract, since they do not depend on the courts to enforce the agreement.

This story contains several familiar lessons. First, it illustrates an old lesson about the problem of cooperation. Even though two people would obtain mutual gains by cooperating, they will not be able to cooperate if they must act simultaneously. This problem is called the prisoner's dilemma or the collective action problem. The story actually illustrates a one-sided version of this problem: the firm and the worker would obtain gains if they cooperate, but because the worker does even better by cheating after the firm commits itself, the firm will not commit itself in the first place.

Second, the story suggests but also casts doubt on a possible solution to the problem of collective action. This solution is the government. If Person 1 and Person 2 enter a contract and expect that the state will sufficiently punish anyone who breaks the contract, then the expected cost of cheating will be higher than the expected cost of cooperation. The parties will cooperate; the prisoner's dilemma is solved. But, as the story shows, legal proceedings are costly and clumsy, so people cannot rely on the law to solve day-to-day cooperative problems. Indeed, that people do not rely on the law to solve day-to-day cooperative problems is clear from both formal research (for example, Macaulay 1963, Ellickson 1991, Bernstein 1992) and casual empiricism. But if the government does not solve these problems, what does?

Third, the story suggests a tentative answer to this question. Sometimes people keep their promises not because they fear being sued, but because they fear developing a bad reputation. If they develop a bad reputation, it will be harder for them to find work and to obtain other good things in the future. So if they value future gains sufficiently, they will be deterred from cheating in the present.

Fourth, the story brings out the problem of character and private information. If the worker is honest, the firm may be able to trust him, but the firm has no way of confirming the worker's claim that he is honest, and the worker has no way of reliably signaling his honesty to the firm. Other kinds of private information, including the extent to which the worker cares about his well-being in the future, as opposed to the present, raise similar problems of verification and signaling. Although the firm and the worker in our story end up working around these problems, the pressure to signal that one has a desirable character leads in other contexts to conformist behavior and the generation of social norms.

Consider some ways in which the worker can *reduce* the chances that he will be hired by the firm. If he shows up at the interview wearing dirty or inappropriate clothes; if he speaks in an impolite or disrespectful or insufficiently deferential way; if he includes in his resume falsehoods, even trivial falsehoods, that are detected; if he expresses weird opinions even if unrelated to the position, or if he expresses *any* opinions that are not germane to the position; if he discloses his membership in a group that espouses idiosyncratic beliefs—if he does any of these things, he will reduce his chances of success despite the tenuousness of the connection between these actions and the requirements of the position for which he applies. Failure to conform to relevant social norms raises suspicions about one's character and reliability in relationships of trust, even when there is no direct relationship between the deviant behavior and the requirements of the job.

The strategic difficulty faced by the firm and the worker is an instance of the general problem of cooperation. The problem is that although cooperation produces mutual gains for those who participate, in the absence of an enforcement mechanism cooperation will not be possible. I use the term "enforcement mechanism" in a special sense. A person who is rational in the economist's most stripped-down version of that term is a person who has complete, consistent, and transitive preferences over a range of options. In certain circumstances that person cannot cooperate with other people. When one relaxes some of the assumptions about human motivation, or one fills in some institutional detail, cooperation becomes possible. When it does, the new characteristics of motivation or details of institution that enable this cooperation constitute the "enforcement mechanism."

In order to be perfectly clear about this point, let me return to the prisoner's dilemma. The following table provides the payoffs of the game, played by Row and Column, each of whom can choose between two strategies, Cooperate and Defect.

	Cooperate	Defect
Cooperate	2, 2	0, 3
Defect	3, 0	1, 1

Row thinks in the following way: "Suppose Column plans to cooperate. Then if I cooperate I obtain 2, whereas if I defect, I obtain 3. Suppose Column plans to defect. Then if I cooperate I obtain 0, whereas if I defect, I obtain 1. So whether or not Column cooperates, I always do better if I defect. Therefore, I will defect." Column reasons similarly, and both players defect. Each would do better if both did not defect, but the optimal outcome does not prevail. The conclusion can be extended to cases involving an indefinitely large number of people who try to cooperate for the purpose of obtaining mutual gains—for example, a family or a union or a political party.

The significance of the prisoner's dilemma has been obscured by some quarrels about the extent of its applicability. First, experiments and intuition suggest that people do not invariably jointly defect in prisoner's dilemmas (Kagel and Roth 1995). But experiments and intuition also suggest that people do not invariably cooperate in prisoner's dilemmas. The point of this thought experiment is to lay bare the incentives that interfere with the perfect cooperation that would turn our flawed world into a utopian one. Having laid bare those incentives, one can proceed to investigate the mechanisms that enable cooperation where it would not otherwise exist.

Second, it is sometimes argued that the highly stylized conditions of the prisoner's dilemma rarely exist in the world. Notice that the terms of the thought experiment forbid the players to communicate in advance or to interact after the game ends, but in reality these conditions are generally violated. Some commentators argue that coordination games (more on this later) are more common than prisoner's dilemmas. But every time one wonders whether one can get away with not reporting income for tax purposes, littering on the street, smoking in a crowded room, driving after a few glasses of wine, gossiping behind the back of a friend, betraying an institution, or breaking a promise, one feels the pull of the logic behind the prisoner's dilemma. That one may not do some or all of these things is due to the existence of enforcement mechanisms, to which I now turn my attention.

Some enforcement mechanisms are institutional and others are psychological or physiological. The latter will be discussed in the next chapter. Institutional enforcement mechanisms can be divided into legal and nonlegal mechanisms. The law encourages cooperation in many ways. For example, it allows people to enter contracts and obtain damages from anyone who breaches a contract. Suppose Row and Column enter a contract. If Row subsequently defects, the state will penalize him with a sanction of 2. Now Row does better if he cooperates than if he defects, regardless of whether Column cooperates. If Column faces the same set of incentives, both players cooperate and the optimal outcome is achieved.

This story is too easy for the reasons suggested earlier. The government is a clumsy tool. Police officers, prosecutors, judges, and juries generally can obtain only a crude, third-hand account of events. Lawsuits are expensive. If the court system cannot distinguish cooperation from defection with any accuracy, and it is costly to use, people will not rely on it for ensuring cooperation. Indeed, most people do not know much about the law, do not allow what they do know about it to influence much in their relations with other people, and do not sue each other when they have disputes.

An explanation for *nonlegal* cooperation begins with the observation that people who defect suffer injury to their reputations. If a person develops a bad reputation, then people will not cooperate with him in the future. Since cooperation is valuable, if a person cares enough about the future he will not defect in the present. But what is a reputation, and how does it get created?

The Repeated Game

Suppose that Row and Column expect to play the game in Figure 1 for an indefinitely long period of time. By this, I mean that there is a low probability at any round that an exogenous shift in circumstances would cause Row or Column to prefer an outside opportunity to the payoff from continued cooperation. Row and Column might, for example, be a buyer and a seller who have jointly invested in assets specific to the relationship, and expect that in the normal run of things the buyer will not stop making orders. Suppose that each player expects that if he cheats in one round, the other player will respond by cheating in the following round. Then as long as each player cares enough about his payoffs in future rounds—that is, he has a low discount rate—he will cooperate rather than defect in each round. As a quick example, suppose that Row's discount rate is 0. If he cheats in round 1, then Column will defect in round 2. If Row cheats in round 2, Column will defect in round 3 as well. Suppose, then, Row cheats in three rounds: then, he will obtain a total payoff of 5 (3 + 1 + 1). If he cheats in the first round, cooperates in the second, and cooperates in the third, he obtains a total payoff of 5 (3 + 0 + 2). If he cooperates in all three rounds, he obtains a total payoff of 6 (2 + 2 + 2), so cooperation is the best move. To be sure, if he cooperates in the first two rounds and cheats in the third, he will obtain a payoff of 7, but then one must extend the analysis to later rounds, and continuing with the same logic shows that the optimal move is always to cooperate. The result can be seen most easily by comparing the stream of payoffs that Row or Column would expect. Cheat on the first round and thereafter, then he can expect to receive {3, 1, 1, 1, . . . }; cooperate on the first round, then at best he can expect to receive {2, 2, 2, 2, . . . }.

The logic extends to games involving more than two players. One should distinguish n-person games in which players randomly meet in pairwise encounters, either cooperate or defect, then move on to other encounters with different people within the group; and games in which everyone must cooperate for the production of some collective good. An example of the first is a sales contract between two merchants who belong to a trade association: each merchant expects the next contract to be with some other merchant. An example of the second is the cleanliness of the community in which no one litters or the safety of a community in which everyone contributes to defense. But the logic of the two examples is the same. As long as everyone has a low discount rate, has sufficient information about the past activities of everyone else, and adopts a sufficiently cooperative strategy, cooperation is possible.

This argument is well-known because it shows that the prisoner's dilemma does not necessarily defeat cooperation. It shows that a few reasonable assumptions about the context in which cooperation occurs converts a prisoner's dilemma into a game in which cooperation is a natural outcome. I say that these assumptions are reasonable because they accord with intuition and experience: people are more likely to cooperate when they expect to have repeated dealings with each other than when they expect never to see each other again. And we will see that many institutions, both government-created and privately created, enhance cooperation by replacing the conditions of the prisoner's dilemma with conditions that promote cooperation, in particular, conditions that facilitate the development of reputation. In this model, reputation refers to people's beliefs about one's history as a cooperator or defector.

This argument, however, suffers from several problems. First, it assumes that the parties correctly interpret each other's actions. This is not always the case. In two-person games Row and Column might agree on a certain quality level for Column's products, then disagree about the quality of the products that Column delivers. As the number of players rises, it becomes decreasingly likely that informal observation, learning, and gossip will be sufficient to keep track of the defectors. Although perfect information is not necessary for partial and even complete cooperation—outcomes might be reliable proxies for moves, error may cause only partial defection rather than complete unraveling—the demands on information are high.[1]

Second, the argument assumes that players will choose the proper strategies, and believe that others will choose the proper strategies. If Row believes (correctly or not) that Column will always defect, then Row will always defect. Even if Column chooses a relatively cooperative strategy that requires defection only for one round after Row defects, Row's strategy will result in both parties always defecting. As P. T. Barnum said of a country store that he managed, "The customers cheated us in their fabrics, we cheated the customers with our goods. Each party expected to be cheated, if it was possible" (quoted in Greenberg 1996, p. 11). Computer studies suggest that the optimal strategy might be to defect only after a pair of defections, as a way of reducing the variance caused by noise (Wu and Axelrod 1995), but in any context there will be an indefinitely large set of strategies that might seem reasonable to a player. In n-person games it is necessary that (1) parties either communicate among themselves (with little error) the outcomes of rounds or observe them directly, (2) parties remember a person's history as a cooperator and defector, and (3) parties adopt extreme strategies, such as defecting perpetually after observing a single defection (Kandori 1992). It is not clear that one should expect players to realize that these strategies are appropriate; indeed, they do not appear to correspond to real world behavior. The general point is that the models provide a case for the *possibility* of cooperation (in the face of the prisoner's dilemma's case for the impossibility of cooperation), but do not guarantee that cooperation will occur or even that the likelihood of cooperation is high.

Third, as mentioned, the argument assumes that the parties have sufficiently low discount rates, when, in fact, some people may have high discount rates. People with high discount rates will not cooperate or at best will achieve a low level of cooperation, depending on the situation. People with low discount rates might not be able to cooperate even with each other, if they do not know whether a partner has a low or high discount rate. Information asymmetries may defeat cooperation.

Finally, the argument depends, of course, on the actual payoffs. If mutual cooperation is highly attractive relative to mutual defection, it is more likely to occur than if mutual cooperation is only slightly superior to mutual defection. In an n-person game, some people may have higher payoffs than other people. As I discuss later, certain people ("high-status" people) might be in such demand generally that they can cheat more frequently than can other people.[2]

The success of a repeated game model in explaining cooperation depends on the parties having extensive information about past behavior and current actions. Sometimes, this assumption does not seem too strong, but often it does. For example, the repeated game theory provides a more powerful explanation for why two merchants involved in a continuing buyer-seller relationship do not shirk on quality (on the seller's side) or delay payment (on the buyer's side), than for why people vote in elections. But even for the case of the merchants, the repeated game theory does not explain all kinds of important aspects of their relationship: why, for example, merchants often become friends, send each other gifts, take each other out to dinner, and participate in the shunning of those who violate an obligation against a third party. More generally, why is cooperative behavior so often surrounded by social norms that require people to shake hands, exchange gifts, conform to clothing fashions, express mainstream opinions, shun people of the "wrong" type? These actions, like delivering a widget on time, are costly; but, unlike delivering a widget on time, they do not in any obvious way produce benefits greater than the costs. The repeated game theory explains why people engage in actions which produce joint benefits greater than private costs in two-person relationships, and provides a suggestive but weak account of cooperation among members of larger groups. A more complete account would strengthen the description of group interaction and explain conformity to social norms.

Signaling

To achieve this more complete account, one must complicate the repeated game model. Suppose that people belong to two "types." Good (or "cooperative") types have low discount rates and bad (or "opportunistic") types have high discount rates. Holding everything else equal, a good type is more likely to cooperate in a repeated prisoner's dilemma than a bad type is, because the good type cares more about the future payoffs that are lost if cooperation fails. The model does not assume that good types have a "taste" for cooperation or are altruistic. Neither a good type nor a bad type cooperates in a one-shot prisoner's dilemma. The model assumes that a person knows his own type but does not know the types of others; he does, however, know the fraction of types in the population, and thus the odds that a potential cooperative partner belongs to one type rather than the other.

The concept of type could be expanded in various ways. The good type might be more likely to choose a nice strategy (for example, punish the defector by cheating in the next round only) over a nasty strategy (for example, punish the defector by cheating in all further rounds). The former strategy is more conducive to cooperation than the latter strategy in some circumstances (Axelrod 1984). Or, the good type might have a better memory about the actions of others, or more accurately interpret the actions of others, or enjoy gossip more than other people do. Because all these ideas can be collapsed into the definition of discount rate, for expository clarity I will generally confine myself to that variable.[3]

The division of people into types is justified by ordinary economic assumptions. It is reasonable to suppose that preferences with respect to the value of future payoffs are distributed among people just like preferences with respect to ordinary goods and services. The division of people into *two* types is a methodological convenience. Types are distributed continuously, and where the two-type assumption leads to conclusions that differ from the continuous-type assumption, I will depart from the former.

The consequences of the assumptions about information are familiar from the literature on signaling (Spence 1974).[4] For concreteness, imagine a group of merchants who deal in similar goods. Each merchant seeks to have a commercial "relationship" with another merchant, by which I mean he seeks to have repeat transactions for an indefinitely long period with a single other person. The merchant's problem, then, is to choose the best partner from the pool of possible partners. Because repeated prisoner's dilemmas are most likely to yield the cooperative outcome when parties have low discount rates, each merchant wants to find a partner who has a low discount rate. Suppose reasonably that the discount rates of different merchants vary: some merchants have invested for the long term, others are little more than con artists out to find a partner and cheat him. The former are good types and the latter are bad types. The good types prefer to match up with each other, and to avoid the bad types. The bad types prefer matching up with the good types, and have little or no desire to match with other bad types.

To distinguish themselves from bad types, good types engage in actions that are called "signals." Signals reveal type if only the good types, and not the bad types, can afford to send them, and everyone knows this. Because a good type is a person who values future returns more than a bad type does, one signal is to incur large, observable costs prior to entering a relationship. For example, if a good type values a future payoff of 10 at a 10 percent discount and a bad type values the same payoff at a 30 percent discount, the good type can distinguish himself by incurring an otherwise uncompensated cost of 8, which is less than the good type's discounted payoff (9) and greater than the bad type's discounted payoff (7). Because the recipient of the signal realizes that only the good type could afford 8, the recipient is willing to enter the relationship. If the original sender lacks information about the type of the receiver, then he or she will demand that the original recipient also send a signal. In equilibrium all the good types send the signal and match up with each other, and the bad types do not send the signal and either match up with each other or not at all. Such an equilibrium is called a "separating equilibrium."

Signals do not always result in a separating equilibrium. Sometimes, an action that served to separate types at time 1 will, because of an exogenous shift in costs, fail to separate them at time 2. If the cost of the signal falls, bad types might join in (they "pool"), in the hope that good types will infer that they (the bad types) are in fact good; or good types will stop sending the signal, because they realize that the bad types can join in, and thus observers cannot distinguish the good from the bad on the basis of who sends the signal. Worse yet, identical sets of parameters will support multiple equilibriums. A signal that separates types at one time might fail to separate types at another time, even though all relevant parameters are the same. Depending on assumptions about equilibrium concepts, it is possible that both good and bad types will send costly signals, even when cheaper signals would produce separating or pooling equilibriums in which everyone is better off. The reason is that given that everyone sends the signal, receivers will believe that anyone who fails to send the signal belongs to the bad type. To avoid this inference, neither good nor bad types will individually deviate from the equilibrium even though everyone would be better off if everyone did. And when signals cost too much relative to the gains from cooperation, and receivers do not believe that only good types send signals, no one will bother sending the signal.[5]

I should note a few complications. In the two-type model, and also in a continuous-type model, "partial pooling" equilibriums can result, in which people of one type randomize among actions. In continuous-type models one might find better types all sending the signals and worse types not sending the signals. In addition, in both two-type and continuous-type models it is more realistic to assume that *signals* are continuous, so that the better types might invest more in a signal while worse types invest less in a signal. Thus, in more complicated models one might find all of the best types sending an identical signal, while progressively worse types shade their actions—they might generally follow their intrinsic preferences while investing a small amount in the signal (Bernheim 1994). Even a fairly bad type finds this investment worthwhile because it prevents others from believing that he belongs to the worst type. Finally, one should assume that having made a prediction about whether a potential partner is a good, medium, or bad type, a person will then engage in a high, medium, or low level of cooperation (or no cooperation at all), rather than assuming an all-or-nothing decision.

To return to the merchant example, the good types can distinguish themselves in two ways. First, as we saw in our discussion of the repeated game model, they can each develop a reputation as a person who rarely or never cheats. They do this, of course, by not cheating in previous games. But this strategy succeeds only to the extent that information flows sufficiently freely and memories are good; and also the strategy obviously cannot be used by a new entrant to the market.

Second, the merchant might send signals. He might, for example, invest in expensive quarters—the way banks used to invest in elaborate buildings modeled after Greek temples, and churches in astonishing cathedrals. This strategy assumes that potential cooperative partners will believe that only a good type will invest in expensive quarters, because only good types can reap high enough future payoffs to recover their costs. Bad types do not invest in expensive quarters, because the discounted value of future returns is not high enough to compensate them for the initial investment. Good types also signal by being business-like, in contrast to bad types, who fail to plan, are poorly organized, are chronically late, and are likely to be obese or to be addicted to alcohol or drugs. Conspicuous consumption—in the form of elegant clothes and jewelry, a luxury car, a redundant staff—can also serve as a signal that one belongs to the good type, although, as we shall see in a later chapter, because demonstrations of cultural competence can also serve as signals, vulgar or inappropriate conspicuous consumption can reveal that one belongs to the bad type. Indeed, because consumption often reveals a high discount rate, exposing as it does the consumer's inability to defer gratification, good types must engage in *conspicuous* consumption, consumption that is stylized to show that it is unlikely to satisfy intrinsic preferences. Whatever the forms of the signal, an appropriately costly signal can prevail in equilibrium as long as observers believe that anyone who sends the signal belongs to the good type. How that belief might be formed is a separate issue, to which I will turn shortly.[6]

Drawing back from the model, one can see that signaling is an important way not only of entering relationships, but also of maintaining them. In earlier times the merchant entering a foreign land for the first time was likely to distribute gifts to the ruler, as a way of showing his commitment to a long-term relationship (rather than, say, his intention to raid); but he was also likely to continue to distribute gifts over the course of the relationship, as a way of showing that he did not expect it to end any time soon. Because outside influences are always changing, one always wants to reassure one's cooperative partner that one remains a good type. And more generally, signaling occurs not just when a single merchant is seeking a cooperative partnership from another merchant, as in the example, but even when merchants in a closely-knit group trade off partners after each transaction, or simultaneously look for people to enter long-term and short-term transactions, or engage in multiple transactions of different lengths. One wants a general reputation as a "cooperator," a person with a low discount rate, and one establishes that reputation both by declining to cheat in repeated games and by sending signals at every opportunity.

But signals do not work perfectly or even particularly well. We will look at these problems in more detail later, but the main problems can be mentioned here. Since people do not have uniform tastes, it is cheaper for some people to send signals than for other people, holding discount rates constant. One signal that will be discussed frequently in this book is that of gift-giving

(Camerer 1988). Gifts can serve as signals if they are costly—either in terms of money or in terms of the time necessary to choose a gift that suits the taste of the recipient—and if they do not benefit the recipient as much as they cost the donor. I will defend this claim in Chapter 4; for now assume that it is correct. Although gifts are natural signals in many circumstances, they reveal the problems with signaling. The problem here is that because altruists can give gifts more cheaply than selfish people can, and wealthy people can afford gifts more easily than poorer people can, a donation can be an ambiguous signal. A generous donation is a function of altruism and wealth, as well as of discount rate. Thus, to interpret a signal properly one must have information about these other characteristics—information that is not forthcoming in many contexts, though it may be in others. When the signal consists of conformity to manners, clothing styles, and linguistic trends, people are vulnerable to the con artist, who exploits people because his unusual skills and idiosyncratic tastes enable him to mimic signals more cheaply than ordinary bad types. Finally, barriers to information conceal changes in costs, which lead people to pool, so behavioral regularities persist even though all people, not just the good types, engage in them. The struggle by good types to differentiate themselves from bad types, as costs change and as bad types hit upon cheap ways of mimicking, accounts for the variation of social norms across time and place.

A large amount of social, family, political, and business behavior can be understood in terms of signals. A signal can be any costly action that enables separation, or at least enabled separation in the past or might plausibly be expected to cause separation in the present. A few brief examples from social, business, and family life will illustrate the argument.

Our social lives are highly regimented by manners. One must hold knife and fork in a certain way, eat with one's mouth closed, look people in the eye but not too much, wear clean clothes and the right sort of clothes, shake hands when introduced, comb or brush one's hair, speak without a regional accent, send holiday cards to one's friends and associates, cover one's mouth when one yawns, avoid staring at defects in other people's appearance, cheerfully say "fine" when asked about one's health, feign interest in other people's stories about their vacations, and apologize when one fails to do any of the above. Some people pay for elaborate and time-consuming hairstyling, invest in contact lenses rather than glasses, pierce ears and other body parts, tattoo or scar or brand themselves.[7] The common elements are (1) the requirement of self-restraint against impulse or inclination, which is "costly" (in time, money, or physical discomfort), (2) the restriction to appearance and observable behaviors, and (3) the arbitrariness of the behaviors. All these elements follow from the signaling model, according to which signals are costly and observable actions with no necessary or intrinsic connection to the beliefs that they provoke. As to this latter point, no reason exists why a fork should be held in one way or the other; people attach significance to how one holds a fork only because of prior and contingent beliefs about the type of person who holds a fork in a certain way. This is why all of the norms listed above differ from country to country, from region from region, even from neighborhood to neighborhood.

It is costly to respect manners; the benefit is cooperation from observers. Some people fail to respect manners because they do not care to incur these benefits. Other people fail to respect manners because the cost to them of respecting manners is idiosyncratically high. The first group of people offends and annoys us not because we derive intrinsic pleasure from observing them hold a fork in the right way, but because their failure to do so shows that they do not value us. The second group of people evokes our sympathy if we discover the source of their failing. A person with a broken hand might hold a fork wrong, as might a person from a foreign culture. We discount the significance of their violation of the convention and look for other signals of their type, and they duly provide elaborate apologies and explanations.[8] But if we cannot discover their idiosyncratic cost, we classify them in the first group and assume that they belong to the bad type. Fearing such a fate for their children, parents train children in manners in order to ensure that their costs are low. Over time this practice muddies the signal, because it means that bad types with proper breeding signal as well as good types with mediocre breeding, so observers discount the signal, and better types (and the wealthier or more dexterous) substitute more elaborate manners as a way of revealing that they belong to the good type. Thus do manners change over time.[9]

Business norms are as elaborate and as fraught with meaning and consequence as social norms. Employees, of course, must observe the same table manners at work as they do at home. But one also finds a variety of practices whose expensiveness and sheer disconnectedness with business are, at first sight, strikingly at variance with the corporate world's devotion to profits. Sales agents give turkeys to their clients on Thanksgiving (why not a discount, instead?), and CEO's take employees to ball games (why not raise their wages, instead?). Bosses must give flowers and other gifts to secretaries on Secretaries' Day, and gold watches to long-time employees when they retire. Firms throw Christmas and birthday parties, and parties to celebrate a deal. People must wear proper attire even during "casual days" when comfort continues to be sacrificed to display (this institution arouses anxiety because people do not know which clothes are "appropriate"). People must use the proper form of address to superiors, subordinates, and equals—never too formal or too casual. And careful attention must be paid to the size of offices and the quality of other amenities. To introduce business as the first topic at a business lunch is a breach of decorum, as is a failure to introduce it at all. Chapter 9 analyzes these norms in detail, so I will not discuss them here, other than to point out that, like social manners, they all involve restraint of impulse or inclination, they all involve observable behavior, they all are artificial—and they all illustrate the use of signals to reveal that one belongs to the good type.

Family norms provide further illustration. Courtship norms, which require exclusive attention to the object of desire, impose costs on the suitor because he must forgo alternative opportunities, much as in business. Gifts circulate between lovers, and among family members, just as they circulate among purchasing agents and clients. The exchange of vows at a wedding attended by friends and relatives is like the exchange of vows when two firms publicly merge. There is nothing surprising about these parallels. When individuals seek mates, they must show that they will be reliable spouses, and after individuals marry, they must continue to show each other that trust is not misplaced. Forgetting a birthday or anniversary creates anxiety that the relationship is on the decline, and calls for a complicated ritual of mutual reassurance. People send signals in order to establish and enhance trust within valuable relationships. Because relationships are as important to people in business and social life as in the family, parallel signaling behaviors arise.

As noted above, social norms are always about *observed* behavior. Norms govern what clothes one may wear but not the order in which one puts on one's clothes. Relatedly, people are more willing to tolerate the violation of norms when the agent tries to conceal his behavior than when norms are openly violated, publicly flouted. People who knew about his homosexuality treated Oscar Wilde far more harshly after he openly acknowledged it (compare Ayres and Nalebuff 1997, pp. 1665–66, n.49). This attitude is common in repressive societies, and is often thought to be paradoxical or hypocritical, but it follows from the logic of signaling. The person who is ashamed by his behavior and takes steps to conceal his conduct from a wider audience incurs extra costs, and by incurring these costs he signals to those who know about the behavior that he belongs to the good type. The person is demonstrating that his discount rate is low but his preferences are idiosyncratic. People do not shun him because he belongs to the good type. When he stops trying to conceal his behavior, people shun him because he is showing that he does not care what they think about him, and this can only be because he discounts the future greatly or prefers to interact with others.

The signaling theory suggests that any costly action can be a signal, that is, a mechanism for establishing or preserving one's reputation. If this claim seems too strong, try to think of a kind of behavior that is truly arbitrary, and chances are that somewhere that behavior has become a signal. I once suggested in an essay that the side of the head on which one parts one's hair is arbitrary, and the editor of the volume told me that when he was a child in England, boys parted their hair on one side, girls on the other side, and any boy who deviated from this norm would be instantaneously ostracized as a "sissy."

This story suggests why anything can be a signal. Sociologists distinguish "statistical norms" from "moral norms," the former referring to patterns of behavior that have no moral content. Among American law professors today, if not among British schoolchildren earlier in this century, it is true that the side on which one parts one's hair has no moral content. But when a pattern of behavior arises completely at random, any (random) deviation from the (statistical) norm becomes salient, and an opportunity for anyone to signal that he belongs to the good type.[10] We might imagine that at time 0, most male and female schoolchildren in some classroom just *happened* to part their hair on opposite sides of the head, but no moral significance was attached to this practice. At time 1, an entrepreneurial kid shows the others that he is a good type because he is able to incur the cost of rejecting a child whose hair deviated from the male norm. Rejecting or shunning another person is costly because one cuts off opportunities for cooperative gains and risks retaliation. If others mimic the leader in order to avoid being labeled bad types, a moral norm would emerge from the statistical norm. People do not engage merely in the behavioral regularity of avoiding the outcast (which they might do independently simply because they do not like him); they engage in the additional behavioral regularity of shunning those who deviate from the first regularity. This is an extremely common occurrence among schoolchildren: anyone who deviates in appearance or behavior—the girl with the funny haircut or the boy who shows up in peculiar clothes or anyone with a physical deformity (Goffman 1963)—presents an opportunity for others to demonstrate their type. What is powerful about the signaling theory is that it shows why schoolchildren and the rest of us devote so much energy and worry to what always seem in the grand scheme of things to be trivial—clothes, hygiene, appearance, manners, forms of speech, and all the other attributes which, because of their salience, present opportunities for others to discriminate against us.[11] The reader should be reminded that a "good" or "bad" type is not necessarily a good or bad person; the label refers to the beliefs of those *within* the group about the hidden characteristics of others.

It should be clear that there are actually two ways in which signaling results in social norms. First, people engage in *costly* actions, like gift-giving, consumption of expensive goods, and shunning of people with certain characteristics, to signal that they value future payoffs more than bad types do. Second, people engage even in *cheap* actions, like combing their hair in one way rather than another, because their deviation from the norm will be punished by *others* who seek to signal their types by taking the costly action of shunning people who act in an unusual way. What we call a social norm, when, for example, we advise well-meaning strangers about how to behave in our communities, is simply a description of the behavior that emerges in these signaling equilibriums. In this model the social norm has no independent power, it is not an exogenous force, it is not internalized; it is a term for behavioral regularities that emerge as people interact with each other in pursuit of their everyday interests.[12]

A few examples might make this argument clearer. If people style their hair in a way that reflects their intrinsic preferences and not concern for what others think, then a distribution of hairstyles should result. The comfort of a particular haircut depends on the shape of one's head, the thickness of one's hair, and other personal characteristics. Since these characteristics vary continuously among members of the population, so should their hairstyles. If people choose hairstyles based on a concern about reputation, then hairstyles will become more regular. Movie stars, fashion models, and other people whose function is to be observed supply the focal hairstyles. Ordinary people mimic them, and in an effort to forestall inferences that their hairstyle converges with the norm by chance (which is costless) rather than by design (which is expensive), people must conform with the models exactly. Differences between hairstyles becomes exaggerated, and the distribution becomes discontinuous. Hairstyles comprise a distinctive behavioral regularity. The regularity reveals to the perceptive observer that the particular behavior is driven by reputation, not by taste.

Another example can be taken from the market. Different sellers usually charge the same price for identical goods: should this behavioral regularity be understood as conformity to a social norm? The answer is no. The behavioral regularity results not from concerns about reputation, but from the dynamics of the market. Sellers who charge high prices will lose their customers, and sellers who charge low prices will not cover costs; both will vanish from the market. However, sometimes prices do reflect conformity to social norms. A business might keep prices for kerosene down after a hurricane, even though this would result in a shortage, because it fears that customers would infer from a high price that the business is an opportunist, a bad type, which cannot be trusted even under ordinary circumstances, when it makes representations about the quality of its products.

An action must be costly in order to be a signal, but this necessary condition is not also a sufficient condition. A further requirement is that observers have an adequate understanding of the cost of an action relative to the gains for different types from cooperation, in any given context, so that they can tell that bad types cannot or usually cannot engage in that action. Thus, it is not necessary for a person to intend an action to be a signal for it to be perceived as a signal (Spence 1975). This means that a person might engage in an action for independent reasons and have no desire to form a relationship with an observer who interprets the action as a signal. This ambiguity is a source of constant tension in daily life. As mentioned earlier, a gift may be perceived as an altruistic display or it may be perceived as a signal of a desire to enter a relationship. If the recipient makes a mistake, embarrassment results. A lawyer might give a colleague flowers in order to thank her for her hard work on a joint project; but if he happens to give them to her on Valentine's Day, she might infer that he seeks an intimate relationship. People like this lawyer have a tin ear for social norms, and he is, in a way, demonstrating that he is a bad type since he shows that he has not invested in cultural competence. An instance of conspicuous consumption—the purchase of a yacht, for example—might reflect the intrinsic tastes of one buyer, given his wealth, while it reflects the effort of another buyer to signal his type. So signals can operate effectively only when people have a rough idea of the tastes and endowments of each other. The fear that others might misinterpret their actions causes people to send signals whose meaning has become clear over time, such as the ritual exchange of presents during holidays, and to consult etiquette books, and to develop protocols. The fear of misinterpretation causes people to exaggerate their conformity to the behavior of others, as the hairstyle example suggested. But to the extent that everyone can copy a ritual or consult an etiquette book, the behavior that results is relatively cheap, the rituals become empty, and the signals muddy.[13] An etiquette book becomes obsolete as soon as it is published. The well-known etiquette book by Post (1992), first published in 1922, is in its 16th edition.

The theory explains a mild paradox of social life: that the ornamental is essential. Ornaments of clothing, hygiene, behavior, speech, even physique, are pervasive, and people who overinvest in them, the dandies, are frequently disparaged. But those who refuse to invest in them are shunned. The question of why people both invest heavily in signaling and deny that they do so is addressed in Chapter 11.

Violation, Hierarchy, and Conflict

People often violate social norms. Sometimes these violations are only apparent, and sometimes they are real. I distinguish four cases.

First, people violate social norms simply because they care more about the intrinsic benefits they obtain from a behavior than about the reputational consequences. They might put little weight on reputational consequences because they are bad types and discount the future heavily. Or they might put enormous weight on the intrinsic benefits: these people lie far from the mean of the distribution of tastes. To illustrate these points, suppose that in a community people ostracize those who trade or socialize with members of an ethnic minority. A person might endure this ostracism and socialize with members of an ethnic minority, either because he does not care about dealing with other people in the dominant community (he discounts the future) or because he values a great deal his dealings with the people in the ethnic minority. A person might dress poorly because he does not care what other people think about him, or because he finds formal dress excruciatingly uncomfortable. A person might refuse to show ordinary respect for the flag because he does not care about lost opportunities from offended patriots or because he strongly believes that patriotism, rightly understood, requires one to abstain from purely symbolic gestures.

Second, people violate social norms because others cannot afford to ostracize them. These norm-breakers are "high-status." Wealthy men sometimes dress expensively, buy costly houses, and treat wives, children, and associates like ornaments or chattel. These behaviors violate common norms against ostentation, waste, and civility. The tycoon can violate these norms simply because other people value dealing with him so much that they will not cut off relations with him even if they do not trust him. For a similar reason, one does not cut off one's relationship with the local gas monopoly when it behaves offensively or incompetently. Movie, pop, and sports stars can flout middle-class norms with impunity because their services are in such high demand. Indeed, as discussed in Chapter 4, because people sometimes infer that those who violate norms must be wealthy and powerful, there is an independent incentive for people to violate norms—as a way of signaling their wealth and power. A common modern form of hubris is for increasingly successful people to violate increasingly important norms until they go too far and are finally shunned by anyone whose cooperation they would find desirable. They violate norms as a challenge; it is a way of saying, You need me more than I need you.

Third, people violate social norms because they belong to groups governed by different norms or because norms change rapidly. Part of the problem is information. Norms of political correctness changed so rapidly that people could be sanctioned for conduct that was considered innocent a few years earlier. People who were sensitive to changes in attitude were able to change their conduct accordingly, but people suffering from tone deafness with respect to these changes did not change their behavior quickly enough and were sanctioned. People violate social norms by mistake when they move from a region with one set of norms to a region with another set of norms, although those in the new region may acknowledge this problem and be forgiving.

Finally, people show their loyalty to each other by ostentatiously violating the norms of a dominant group. A rich vein of examples is found in the anthropology of the American teenager. One way for a teenager to commit himself to other teenagers is to refuse to use the signals necessary to show one's cooperativeness to adults. The teenager who wears a nose ring, acquires a tattoo, wears peculiar clothing, violates the conventions of hygiene, and curses at inappropriate times makes himself unfit as a cooperative partner for adults, but thereby is an excellent cooperative partner for other teenagers. Precisely *because* this teenager cuts himself off from the adult world of (deferred) material reward, he becomes a reliable partner for other teenagers. He is more reliable than the teenager who does not do these things, because ostracism from his gang of friends leaves him no opportunities, whereas a teenager who has no friends can at least interact with adults. But I am not talking just about children. Criminals and members of groups that are rejected by the dominant society develop their own commitment mechanisms, including distinctive dress and behavior (for example, tattoos), for similar reasons. People seeking admission to fraternities and other social clubs commit themselves by secretly violating the norms to which they conform in public. The exemplary case is that of the homoerotic rituals required for admission to homophobic fraternities. The implicit threat to expose the behavior of those who later betray the group discourages free-riding on group activities.

It is important to be clear that teenagers, fraternity members, and even criminals do not necessarily have low discount rates. Teenagers and others who reject mainstream society violate conventional forms of behavior not because they do not care about future payoffs, but because they seek future payoffs from people outside mainstream society. Those who observe these violations refuse to cooperate with this group of people because they believe that they are unreliable, but their unreliability is a result not of their discount rate but of the contempt they hold for the observers. The incentives here are different from the incentives described by the signaling model, and I will henceforth refer to the model described in the prior paragraph as the *commitment model.*[14]

The Norm Entrepreneur

The cooperation game requires that the signal be costly, but nothing about the game dictates the form of the signal. As long as an action is both actually and apparently costly, it can serve as a signal that the sender belongs to the good type. Gift-giving, advertising, the purchase of fashions, and other forms of overt consumption can serve as signals in the cooperation game. But signals can be ambiguous. As noted earlier, giving a gift or attending a parade does not require that a person have a low discount rate if he or she enjoys these activities. The question is, then, how do certain behaviors take on symbolic value?

One answer draws on the idea of focal points, already mentioned in passing. Historical coincidence, physical qualities, and mere deviation from statistical regularities can cause people to associate certain behaviors with certain qualities of character. Days on which great events occur become holidays. People who gain fame through extraordinary action or sheer luck become models that ordinary people imitate. The importance of serendipity in the establishment of signals is proved by the diversity of signals across regions, cultures, and time periods.

Other signals are fabricated. Martin Luther King, Jr. Day, for example, was created by the federal government; it did not arise spontaneously. Once it was created, people could signal their loyalty to the civil rights movement by making gestures of respect on that day, in a publicly visible way. But truly fabricated signals are rare. Usually, when the government or private individuals succeed in establishing certain actions as signals, they do so by drawing the public's attention to one of several conflicting focal points. To show respect for civil rights, should one have a holiday or some other celebration? Should the holiday be in honor of King or another leader? Should it be on King's birth or death or on the date of one of his accomplishments? History supplies a number of focal points; authoritative individuals enable citizens to coordinate around one.

To analyze these points more carefully, it is useful to imagine the cooperation game embedded in a larger game. Prior to the first move of the cooperation game, a "norm entrepreneur"[15] announces that a particular action will be a signal. (In real life, many norm entrepreneurs compete by proposing different signals.) The norm entrepreneur states that a particular action, for example, attending a public ceremony, will be understood as a signal of cooperativeness. The cooperation game is modified in the following way. Every person, in his capacity as a sender of signals, chooses among all possible actions that might serve as signals. The sender might choose the signal recommended by the norm entrepreneur, or might choose another signal. After the cooperation game is played, the norm entrepreneur receives a payoff that is a function of the number of people who issue the signal that he recommends. In some contexts it makes sense to say that the norm entrepreneur receives payoffs if he eliminates a pooling equilibrium in which everyone sends costly signals; in other contexts it makes sense to say that he receives payoffs if he creates a separating or pooling equilibrium. The norm entrepreneur does not, ex ante, have perfect information about the payoffs of the parties in the cooperation game, so he cannot invariably choose the signal that maximizes his payoff. Because the game repeats itself, however, the norm entrepreneur can invent new signals in future rounds if earlier signals fail to yield high payoffs.

In choosing among signals, people have only a rough idea of which signals will allow them to distinguish themselves from bad types and which signals will not. They must also worry about whether others will recognize the signal as the distinctive action of the cooperator, as opposed to simply a pleasurable form of behavior. This is a coordination problem: as long as everyone, or most people, believe that certain behaviors serve as signals, and as long as these behaviors actually have the right cost structure, these behaviors can act as signals. Senders of signals will rely on custom (for example, gift-giving), but also on meanings provided by the Zeitgeist (for example, anti-communism in the 1950s). The suggestions of authoritative norm entrepreneurs shrink the pool of the signals to choose from. As long as enough senders follow the suggestion of a person, he will become a norm entrepreneur. (An important condition, again, is salience; that is why even politically ignorant celebrities can make influential political and cultural announcements and the descendants of royalty, no matter how undistinguished their abilities, could always find a following during crises of succession.)

The significance of the NE game, as I will call it, is that it shows that if a signal fails to create a separating equilibrium or a pooling equilibrium, the norm entrepreneur has an incentive to create a new signal, in the hope that it will be more successful. But it also should be emphasized that signaling can arise spontaneously around random events that become focal, with norm entrepreneurs playing no role. When signaling behavior has a high public profile, it is often referred to as symbolic behavior. I will refer to the creation of new symbols to replace old symbols that fail to produce a desired equilibrium as *symbol transformation.*

People self-consciously try to create new signals or to modify old ones. The incentives to do this arise because of the ambiguity of signals: what is understood as a signal, as opposed to an action taken for independent reasons, is often a historical accident. They also arise because sometimes signaling injures a great many people (for example, racial discrimination or dowry competitions), or simply cause discomfort (formal clothing). Arbiters of taste help define signals in social settings. So do sellers of consumer goods, when they use advertisements to promote a style of life that requires the purchase of their goods. Managers, employees, consultants, and other market actors help define signals in business settings. Politicians and protocol experts help define signals in political settings. The most striking current examples of this phenomenon are the successful efforts by feminists, blacks, and supporters of gay rights to modify the words people use to refer to women (the elimination of "stewardess," "policeman"), blacks (the promotion of "African-American"), and homosexuals (the widespread acceptance of "gay and lesbian").

What motivates the norm entrepreneur? People earn returns when they contribute to shifts in mass behavior. Corporations earn money when consumption of their products becomes a symbol of cultural competence. That is why they invest so much in shifting people's beliefs about the kind of person who wears a certain pair of sneakers, or a blouse with a certain label on it, or a car that has a distinctive hood ornament. Politicians earn the loyalty of vast groups of people when they effect a shift in behavior that benefits these people. Journalists and academics receive all the benefits of fame when their pronouncements influence millions. But norm entrepreneurship is risky precisely because one breaks a norm by challenging it. Few people take this risk, and those who do either have tastes or values that lie on the extremes of the distribution, or else have immense talent and charisma, so people cannot afford to shun them. In politics we observe people with immense talents but ordinary tastes, people with ordinary talents but extreme tastes, and people with both immense talents and extreme tastes. The first group consists of politicians, the second of martyrs, the third of leaders and heroes.

The State

Government actors can serve as norm entrepreneurs. Authoritarian leaders have often succeeded in creating holidays, memorials, and even ideologies about the nation that force people to take a public position that is consistent or inconsistent with goals of the authorities. When people pool around the newly minted signals, the government has succeeded in its norm entrepreneurship. But governments do not always succeed, and governments in democracies often fail because leaders must always satisfy their followers. Nancy Reagan's "Just Say No" campaign against drugs was a failure. As often happens with failed norm entrepreneurship, the result was ridicule, not conformity. Instead of leading, politicians in democracies must conform to signaling equilibriums or risk expulsion from office. The official who argued in favor of communism in the 1950s was not sending the right signal; nor is the official today who says that people should be permitted to burn flags.

From a normative perspective, it will often be desirable for the state to modify signaling equilibriums, but it will often be the case that it cannot. I will say more about this later on, but for now, keep the following distinction in mind. The cooperation game can produce two kinds of collective goods. The *internal* collective good consists of the mutual gains that two people obtain when they match up and cooperate. Notice that the actions that serve as signals are not themselves valuable; they are valuable only insofar as they facilitate cooperation. The *external* collective good consists of any value (positive or negative) that is, as a side effect, produced by the fact that many people engage in identical signaling behavior, whether it be separating or pooling. If the signal is voting, then the external collective good is the generation of information that occurs in elections. If the signal is discrimination against an ethnic group, then the external collective good is bad—it is the injury to members of the group, who are, in a sense, outside the game. It is important to recognize that the existence of one signal—for example, discrimination against blacks—and the existence of another—for example, forms of gift-giving or voting—can be arbitrary, so in theory a bad equilibrium, one that produces external collective bads, can collapse and be replaced with a good equilibrium without the loss of internal collective goods.

One last distinction is useful. When a law changes an equilibrium, it has two separate effects. The first effect is *behavioral:* the law affects the actions people take. If the law taxes an action that has served as a signal, and as a result people engage in that action less frequently and substitute some other action, then the law has influenced their behavior. It has affected the amount of signaling and the kinds of signals that are issued. The second effect is *hermeneutic:* the law changes beliefs that people have. If the taxation of a signal results in the collapse of a separating equilibrium, then people will no longer infer that a person who sends that signal belongs to the good type. Thanks in part to the law, but also to many other developments, people draw different inferences today than they did in the past about the characters of those who engage in racial discrimination. The meaning of racial discrimination has changed. Many other examples of this phenomenon will be supplied throughout the rest of this book. What is important to understand for now is that the law can cause changes in the inferences that people draw about a person on the basis of his actions, and that this function of the law is very important though frequently overlooked.

One should understand that in any given equilibrium some people will do better and others will do worse than in any other equilibrium. If in one equilibrium there are racist social norms, certain whites will do better, while most blacks will do worse, than in another equilibrium, where social norms require tolerance. Individuals and governments thus have powerful incentives to change the equilibrium (or what is the same thing, the social norm) in one direction or the other. When they succeed, internal collective goods may become harder or easier to obtain, and one external collective good or bad may vanish and be replaced by another. Behavior changes as people send new signals, and cooperate more or less; and beliefs change as people develop new associations between the behaviors that others engage in and the characters that they have.

Summary

The following definitions are of terms that play an important role throughout the rest of the book. The terms are all related to nonlegal cooperation, and are defined by reference to the signaling model discussed in this chapter.

Cooperative partner. A cooperative partner is anyone with a sufficiently low discount rate and sufficiently similar interests, such that he and another cooperative partner can overcome the prisoner's dilemma and obtains gains through long-term cooperation. This might occur in family, social, business, and political relations; the ungainly term is used so the discussion can be abstract.

Reputation. When a person defects in a cooperative relationship or fails to send an appropriate signal, these actions are observed by others with some probability and transmitted through gossip. When a person defects or fails to send the appropriate signal, observers will infer that he belongs to the bad type. Reputation consists of inferences that other people draw about one's type, based on such past actions. To be sure, one can have a reputation for being a skilled tennis player or a generous humanitarian. But I will generally use reputation narrowly to refer to beliefs about the more abstract quality of cooperativeness.

Trust. A cooperative relationship is characterized by trust when neither party anticipates that the other party will not defect because of the other party's concern about his own reputation. Thus, trust is defined narrowly, excluding the possibility that one person trusts another to keep his promise because of the threat of legal enforcement.

Solidarity. A group of people has solidarity when everyone places sufficient trust in others that ordinary cooperative relationships can be maintained (in pairwise interactions) and public goods can be produced (in group interactions). (Compare Hechter 1987, p. 18.)

Social norms. Social norms describe the behavioral regularities that occur in equilibrium when people use signals to show that they belong to the good type. Social norms are thus endogenous: they do not *cause* behavior, but are the labels that we attach to behavior that results from other factors. Social norms should be distinguished from behavioral regularities that emerge in cooperative relationships simply because they are value maximizing. For example, in a merchant community the exchange of gifts on holidays reflects a social norm, whereas the rule that the seller pays for freight insurance probably does not, but simply reflects the cost-minimizing strategy.

Tradition. Traditions are social norms that have the sanction of time. They describe a signaling equilibrium, but the signals are less ambiguous than the actions demanded by social norms, just because their age gives them prominence. For example, a social norm demands that I give my secretary presents from time to time to show appreciation; giving a present on Secretaries' Day or Christmas is traditional and thus unlikely to be misinterpreted. Giving a present on Valentine's Day is not traditional and thus may be misinterpreted. The term *culture* will also be used sometimes to capture these ideas, though it implies less of a reliance on the past.

Community. A community is a group that (a) has solidarity and (b) is characterized by a great deal of signaling, reflected in a great many social norms or traditions. A group of strangers who find themselves in a lifeboat may develop a great deal of solidarity over time, since all depend on everyone else for mutual survival. Social norms will no doubt develop rapidly to deal with various problems, but they are not strong enough or old enough (or apparently old enough) that we would call this group a community.

The arguments presented in this chapter are abstract and incomplete. Readers might demand more examples and greater attention to problems with the game-theoretic framework. As to the latter point in particular, readers might complain that there is insufficient discussion of the problem of multiple equilibriums, of alternative strategic problems, of technical complications that arise from the use of continuous types, and so on. Some of these complaints are addressed in succeeding chapters. As to methodological issues that are not fully analyzed, some of these gaps are filled by scholarship cited in the notes, and others reflect mathematical and theoretical problems that have not been solved and to which I have nothing to contribute.

3

Extensions, Objections, and Alternative Theories

The signaling theory of cooperation and social norms can be used to understand and criticize a variety of laws. But one must acknowledge two problems with the theory. The first is that of testability; the second problem is that of underinclusiveness.

Testability. Because signaling models produce multiple equilibriums, it is difficult to use them to generate testable predictions. The number of equilibriums can be reduced by making strong assumptions about what constitutes rational behavior, but these assumptions are controversial, and empirical tests—which are now really joint tests of the model and of the rationality assumptions—are hard to devise and their results are hard to interpret.

Nevertheless, hypotheses derived from the signaling model will be examined informally in light of common intuitions and data. Because the intuitions and data are presented casually, it will be useful here to discuss how one would more formally test the theories. The common theme among the theories is that good types—people with low discount rates—are more likely to conform to social norms than bad types are. To test this proposition, we need to have (1) a measure for discount rates and (2) a measure for conformity to social norms.

The independent variable is discount rate. A person's preference for payoffs sooner rather than later cannot be directly observed, so one must rely on a proxy to measure discount rates. Economists have used the following proxies for a low discount rate in order to test the relationship between discount rate, criminal behavior, and addiction: not having a savings account; smoking; having sex without contraception; being at a young age when one first smoked, drank or had sex (Glaeser 1998, p. 3); being poor; being less educated; and being younger (Becker and Mulligan 1997). Because studies have found associations between low discount rate, crime, and addiction (Glaeser 1998, Becker, Grossman, and Murphy 1991), one can add criminal behavior and addiction as proxies. So one predicts that people with all or many of these characteristics—call them "bad-type characteristics"—are less likely to conform to social norms than people without them.

There are complications. A monk might take a vow of poverty but have no problem deferring gratification. A member of a criminal gang might commit crimes to show his loyalty to the gang, not because he has trouble restraining his impulses. An adequate empirical test would have to control for these possibilities.

The dependent variable will be a measure of conformity to social norms. Conformity to social norms means signaling. So one predicts that people with bad-type characteristics will engage in less signaling than people without these characteristics, holding everything else equal. The trick is holding everything else equal.

The problem is that what might emerge as a conventional signal in any group is to a certain extent arbitrary. Any action will serve the purpose, and multiple actions will serve the purpose, as long as they have the right cost structure, and cost itself does not provide much of a constraint because one can control the cost of many actions (for example, gifts can be more or less expensive). So in one community people might signal by voting, wearing formal dress, praising the government, going to church, and inviting each other out to dinner, while in another community people signal by attending political protests rather than voting, wearing nose rings, condemning the government, smoking, and helping the poor. Indeed, the two communities might be in the same location, the signals of one community might be reactions to the signals of the other community, and there may even be some overlap in membership. So one cannot make the straightforward prediction that people with bad-type characteristics are less likely to vote; good types in the second community do not vote because voting is not recognized as a signal. Instead, one must hold community membership constant; then one can predict, for example, that a person in the first community who has bad-type characteristics will vote less often than a person in the first community who does not have those characteristics.

Further complicating matters, a deviant subgroup might adopt signals that are self-consciously in opposition to the signals used by the majority. Smoking, committing crimes, and other actions that violate the norms of the majority are often used to bind members of the subgroup. In the majority, a person suppresses a desire to smoke or drink in order to signal that he belongs to the good type. Those who have no taste for cigarettes or alcohol are lucky. They benefit from the crudeness of signals, the inevitability of error. In the subgroup, a person suppresses a desire (or moral commitment) *not* to smoke or drink in order to signal that he belongs to the good type. An astute observer can distinguish behaviors that are signals and similar behaviors that satisfy intrinsic preferences. Signaling behavior is more ritualized or stylized, because the person sending the signal wants others to know that his behavior is not simply the indulgence of a taste, and one way to convey this information is to engage in stereotyped behavior rather than idiosyncratic behavior. An example is the difference between social drinking and alcoholism; another is the difference between wearing old pants and wearing new pants that have been manufactured to look old. But these subtle differences, though intuitively understood by everyone, do not usually appear in the data used by the social scientist.

Underinclusiveness. The cooperation theory holds that people will always follow their intrinsic preferences when they are not observed by people who have an incentive to impose sanctions. But this prediction is generally regarded to be false. People appear to act against interest when there is no penalty. Examples include tipping waiters in restaurants to which one never expects to return, stopping at red lights when no one is around, returning wallets found on the street, rescuing strangers from danger, and accepting or rejecting offers on the basis of the fairness of the offer (Thaler 1991). Experimental work suggests that people sometimes cooperate in prisoner's dilemmas and similar games in which cooperation is not an equilibrium outcome (Kagel and Roth 1995).

These problems are occasionally said to undermine rational choice theory, but a more accurate assessment is that rational choice theory can be used to explain some social phenomena but not other social phenomena. It is not clear where the line should be drawn, and it seems unlikely that one can say in the abstract when the theory runs out. Many theories based on rational choice cannot explain important behaviors or do so only by making strong assumptions that render the theory tautologous. But the methodology itself does not fail until it becomes clear that better methodologies are forthcoming. Rational choice theory has some successes, and these successes make it attractive compared to its competitors.

What are these competitors? Sociology seek patterns in social organization that transcend the actions of individuals. Theories derived from cognitive and social psychology locate the source of behavioral regularities in the structure of the brain or the influence of upbringing. These disciplines have produced many interesting results that pose a challenge to rational choice theory, and they have attracted the interest of many law professors. But they have not yet entered the mainstream of legal scholarship. Perhaps they will in the future, but I will leave that to others.

Many rational choice theorists see a third way to address the theory's problems: maintain the general framework of rational choice theory, especially its commitment to methodological individualism, but relax some of its less realistic assumptions. The following sections describe a few of the resulting theories.

Altruism

Altruism can explain cooperation across a range of contexts. The patron tips the waiter in the strange restaurant because the patron cares about the waiter's well-being. A person refrains from littering on the beach because he cares about the feelings of the other bathers. The taxpayer reports all of his income because he cares about the welfare of his fellow citizens. Cooperation in prisoner's dilemma experiments results from players' desires to benefit each other.

Critics argue that the use of altruism in rational choice arguments is tautologous. This is not true; or, more precisely, whether it is true or not depends on how altruism is used in the argument. Take the examples above. Some patrons tip waiters and other apparently identical patrons do not. If the theory assumes that everyone feels altruism toward waiters, then the theory is falsified; if the theory assumes that some people feel altruism toward waiters and some do not, then the theory verges on tautology. A popular theory for why people donate money to charities is that they feel altruism. But this theory is inconsistent with the fact that people do not always reduce their gifts when the charity receives more money from other sources (Sugden 1982). The theory is also inconsistent with the fact that the size of a gift varies with the amount of publicity given to the gift. More examples will be discussed in Chapter 4. The problem with the theory that altruism explains philanthropy is not that it is tautologous, but that it is false.

The problem, I should stress, is not that altruism does not exist. There are persuasive reasons for believing that altruism exists among kin and even members of social groups.[1] But altruism cannot account for collective action as it is usually observed. Altruism cannot account for norm-driven behavioral regularities. For example, it cannot explain the fact that gifts are routinely exchanged on holidays and special occasions, not whenever a transfer would maximize the utility of the donee as filtered through the donor's utility function; or that people give waiters tips of 15 to 20 percent rather than an amount that varies with their altruistic feelings. The proper theoretical approach is not to give altruism any particular explanatory weight in a theory of cooperation, but to treat it as a taste like any other, which gives rise to its own collective action problems that need solution.[2]

Status and Conformity

Another form of interdependent utility is discussed under the labels "status" and "conformity." A person might derive utility from having more wealth, or more of some kind of good, than other people. The possession of goods, or at least of certain status goods, that others do not have confers status on the owner. Or a person might lose utility from having different goods from the goods possessed by everyone else. The failure to be in the same position as others confers disutility. If people care about status, they will be trapped in a prisoner's dilemma, as they overconsume in an effort not to lose ground to others. If people care about conformity, they overconsume or underconsume in an effort to consume what everyone else is consuming.[3] As an example of the problems created by status-seeking, suppose everyone in a community owns a safe car, which is the kind of car that everyone prefers. But people also care about status, and one person buys a fancy but unsafe car because he derives utility from his possession of a car that is more expensive than the cars owned by others. This action confers disutility on the others, so they trade in their safe cars and buy the fancy car. But if everyone has the identical fancy car, no one has status by virtue of owning a car that is fancier than the cars owned by everyone else. The result is that everyone is worse off because they have unsafe cars while being unable to satisfy their taste for status. Frank (1985) explains the deviation of wages from marginal productivity as the result of high-productive workers paying off low-productive workers to accept their lower status in the hierarchy. Bernheim (1994) shows how a concern for status can cause behavioral regularities (and see McAdams 1995, 1997). Akerlof (1997) pursues similar themes under the rubric of "social distance."

Relying on strong assumptions about people's utility functions is risky, because it limits the explanatory power of a theory. Status theories cannot explain why some people seek status and others do not. The cooperation model holds that status is not intrinsically important, but is a label assigned to people who are believed to possess characteristics that others seek in cooperative partners. When I say that people seek status, I mean that they seek the cluster of goods, characteristics, and accomplishments that will make others think that they have very desirable qualities. This theory has testable implications, discussed in Chapter 4.

Herd Behavior

Behavioral regularities can arise in the absence of reputational effects and in the absence of interdependent utility functions. They arise, for example, when people make inferences about the value of options from other people's actions, a phenomenon that may lead to "herd behavior." Suppose that everyone lines up to enter a fashionable restaurant while ignoring the equally good restaurant across the street. A reputational explanation suggests that people go to the fashionable restaurant in order to show that they are good types. An explanation that rests on interdependent utility assumes that people enjoy patronizing restaurants that other people patronize, but it does not explain why this would be so. The herd behavior explanation holds that people have only partial information about the quality of restaurants, and imitate other people in the expectation that inferences based on other people acting on *their* partial information reflect aggregation of information about the quality of restaurants. Thus the social norm—the behavioral regularity of patronizing a certain restaurant—arises as a result of people's incentives to avoid bad outcomes that would occur if they relied on their own partial information (Banerjee 1992; Bikchandani, Hirshleifer, and Welch 1992).

The herd behavior model is similar to the reputation model. Both models assume stripped-down preferences, and derive the incentive to conform from agents' partial information. The main difference is in interpretation. Under the reputation model people engage in behavioral regularities because they have limited information about the tastes of others. Under the herd behavior model people engage in behavior regularities because they have limited information about the value of goods or services. The reputation model seems to provide a truer explanation for social norms as they are commonly understood. One thinks that people conform to social norms because they care about their reputation or fear sanctions. But one does not sanction people who make poor decisions, based on insufficient information, if those decisions do not injure one.

Still, the distinction between behavioral regularities caused by herd behavior and behavioral regularities caused by reputational concerns is not always clear. Herd behavior models are sometimes used to analyze fashion, whose characteristic volatility can be derived from imitative behavior. So it may seem that social norms and norms of fashion are different phenomena. But intuitions are complex and generalization is hazardous. Are high heels or ties required by social norms, or are they mere fashion? Rather than debating semantics, I hold that any behavior regularity that results from partial information—either about the value of some activity or about the character of people who engage in it—is described by a social norm, whereas behavioral regularities that arise because of coincidences in tastes, technologies, and budget constraints (such as wearing warm clothes on a cold day), or because of institutional competition (such as market prices) are *mere* behavioral regularities.

Emotions

Frank (1988) notes that most people have only limited control over certain physiological traits, and most people can, most of the time, correctly interpret people's physiological traits. For example, most people blush when they lie (at least to someone they sympathize with), and most people will interpret the blush to mean that the speaker is lying or is up to no good. Now, if one has the power to control one's physiological traits, one might as well deceive people; one might have a promising career as a con artist or lawyer. Nature provides lots of examples of such deviants in evolutionary games. But deviance by some players is a stable strategy only if most players do not defect.

At first sight Frank's theory seems to add to rational choice theory the assumption that people have a "preference" to cooperate, but his theory is more interesting than that. People cooperate rationally in response to the fact that their body gives them away if they have bad motives. Having such physiological traits, moreover, confers an advantage, because people who cannot violate commitments, or who cannot do so without giving advance warning, are good cooperative partners, so they will be sought out by others for cooperative ventures. One can thus tell an evolutionary story about why in equilibrium many people will have such physiological traits.

Frank argues that his theory explains both the intra-group cooperative behavior that concerned us in Chapter 2 and the relations between strangers. The first claim is plausible, but it has little methodological payoff. I do not break a promise because (1) if I break a promise, the victim and others in my group will henceforth avoid me; and (2) my body will give me away before I break my promise, allowing the victim to take precautions that reduce my payoff from cheating. Given our assumption that a sufficiently dense network of communication exists, the two theories have exactly the same implications. Applying Occam's razor, however, one prefers the first.

Frank's theory has more promise for extra-group cooperative behavior. Suppose I enter a store with the intention of shoplifting. But my heart pounds, my temples throb, and sweat drips down my forehead. The proprietor notices my suspicious behavior and observes me carefully, and accordingly I refrain from shoplifting.

But is this why people tip waiters? Frank does not claim that the customer feels an emotional bond to the waiter after the delightful service, and so cannot bring himself to "cheat" on the relationship. Nor does it seem plausible to say that the customer with bad intentions would blush or otherwise reveal himself as he sat down, so the waiter would discover in advance that this person is a bad type and will not tip him, and so would give him the bad service he deserves.

In fact, Frank's argument is different. Frank suggests that most people tip the waiter in the strange restaurant in order to practice being honest. Character is like a muscle: exercise it frequently and it will be strengthened. By being honest as much as possible, however trivial the circumstance, one develops the ability to resist the temptation to cheat when the payoffs are larger. (Frank has to make the further assumption that people are impulsive.) Given one's body's propensity to give one away, having this ability to resist this temptation is visible to others, and thus one will be regarded as an attractive cooperative partner even for ventures involving the highest stakes.

There is a great deal here that is plausible, but one ought to have reservations. Is the muscle metaphor appropriate? Does one develop one's character by exercising it, or is it more likely that no number of instances of trivial virtue will make it easier for a person to resist cheating when the payoff is high enough? I will discuss these issues in more detail in Chapter 11.[4]

Internalization of Social Norms

One solution to the problem of the waiter in the strange restaurant is to say that people internalize social norms. This idea is often captured by the distinction between shame and guilt. If I tip so that people do not think I am cheap, then I tip to avoid shame. If I tip to avoid the unpleasant sensation that I have done wrong, then I tip to avoid guilt. Fear of shame cannot explain why people tip waiters in strange restaurants, unless one makes the implausible assumption that their action might be detected and publicized, but fear of guilt can.

This argument seems reasonable, but it is methodologically sterile. It is reasonable because people do conform to ethical norms even when no one observes their behavior, and because it takes account of one's interior life, something that reputation models relentlessly ignore. The argument is methodologically sterile, because no well-developed theory of guilt allows us to make predictions about when fear of guilt deters people from engaging in certain actions and when it does not, or what kinds of people feel guilt and what kinds of people do not. So if we observe that people understate their taxes but they do not litter, or if we observe that some people litter but others do not, we cannot rely on a theory of guilt for an explanation. By contrast, the signaling model supports a reputational theory that can in principle explain why people conform to some behavioral regularities and not to others.

It has been suggested that people gradually internalize norms to which they initially adhere for reputational reasons. A person tells the truth because he fears a reputation for dishonesty, but over time he internalizes the norm of truth telling, and tells the truth because if he lied he would feel guilty. I do not know whether this is true, but it certainly sounds reasonable. A better example might be that of a person who migrates from a culture in which people greet by bowing to a culture in which people greet by shaking hands. This person initially shakes hands because he realizes that if he failed to do so, he would offend people. Over time, he gets into the habit of shaking hands. He does not have to think about whether it is appropriate in a particular context; he just does it, instinctively relying on a newly gained cultural competence.

Even if this theory is correct, its implications are limited. It does not predict that some behavioral regularities are more likely to occur than others, because it does not hold that people are more easily habituated in some activities than others. The theory does imply that social norms that arise for reputational reasons might be *sticky,* that they might not collapse as rapidly as they would if people did not feel guilt or were not subject to habituation, but this is not a particularly interesting implication. Most theories of law and social norms that rely on internalization essentially treat social norms as exogenous. Formally, if a social norm exists, these theories assume that people have a "taste" not to violate it. This assumption can lead to some interesting results. Akerlof (1984, ch. 3) produces equilibriums in which people have no incentive to challenge an inefficient status quo. But the approach begs the question of how the relevant social norms arose in the first place, and does not allow one to identify factors that undermine or enhance such social norms. The legal scholarship in this vein is rich in insights but in the end unsatisfying, because it does not explain how law influences social norms.[5]

Bounded Rationality

Standard economic models assume that agents maximize their gains within budget constraints. These models put an unrealistic burden on the memory and analytic powers of the agent: nobody does or can take account of everything when deciding how to act. And once decision costs are taken into account, the models fall into an infinite regress (Conlisk 1996). Although the assumption that people optimize may allow for reasonable approximations of human behavior in some contexts, it does not in other contexts. One response is to assume that people engage in a more or less drone-like mimicking of other people's behavior. The model typically works like this: In a large group of people each person observes some but not all or even many of the actions of others. One might imagine that at the beginning, people choose randomly from a set of strategies; then they interact; then people mimic the behavior of those, from the pool of those whom they observe, who seem to have done well in the interaction. Over time more successful strategies drive out less successful strategies.

The usual example of this model comes from traffic conventions. Suppose that people drive on the roads, but there are no laws or conventions about driving on the left or right. A desirable convention is either always drive on the right or always drive on the left; one assumes that people are indifferent between them. A bad convention is, for example, always drive on the north or west side of the road, since this will cause accidents. Assume that people do not observe anything other than their own experience on the road or perhaps a small subset of other people's experiences; they choose a strategy at random, then follow it until they observe that people following a different strategy avoid accidents more successfully than they do, at which point they mimic the new strategy. Over time the bad strategies will be driven out and one of the good strategies will become universal—it will become a convention (Young 1998a).

A similar approach can be used to explain the evolution of property rights. Suppose two people come upon a valuable piece of property in the state of nature. They can choose between two strategies: pounce aggressively on the property or shyly wait and see. The combined strategies from best to worst payoffs are: (aggressive, passive); (passive, passive); (passive, aggressive); (aggressive, aggressive). The idea is that I get the property if I pounce on it and you do not; if neither of us pounces on it, it begins to rot as we hem and haw; but if we both pounce on it we get in a costly fight. One equilibrium involves randomizing behavior, but the best equilibrium is one in which both sides coordinate by alternating between passiveness and aggression. Scholars have argued that possession might be a focal point around which parties coordinate. If one person possesses the property, then he will protect it aggressively while the other defers. That many animals have similar conventions lends plausibility to its application to humans (Sugden 1986, Hirshleifer 1982, Maynard Smith 1982).

Although this theory is enormously suggestive and has some important normative consequences (see Chapter 10), it is unclear how much behavior can be explained using evolutionary games such as these. It is hard to believe that they explain why people tip in strange restaurants, given that, by hypothesis, no one observes the person's behavior. I have not found many useful applications for the legal issues that are the subject of this book—probably because these issues all concern behavior of which optimization seems to be a better approximation than imitation. However, the evolutionary approach provides a cleaner explanation than the signaling approach does for norms and conventions such as those that govern driving; this is a helpful reminder of the danger of monolithic approaches to explaining cooperative behavior.[6]

A recurrent objection to the theory described in Chapter 2 is that signaling "can't be all that there is." Readers will object that racial discrimination, patriotism, ceremonial gift-giving, clothing fashions, and other complex social phenomena discussed in subsequent chapters involve more than the efforts of members of a group to signal to each other that they belong to the good type. These behaviors bubble forth from a cauldron of instincts, passions, and deeply ingrained cultural attitudes.

My response is that this book reflects a methodological commitment. My claim is that rational choice theory can shed light on social norms by focusing on the reputational source of behavioral regularities to the exclusion of their cognitive and emotional sources. I do not claim that rational choice theory can offer a complete explanation of social norms or of cooperation. Cognition and emotion are not irrelevant. They are just not well enough understood by psychologists to support a theory of social norms, and repeated but puzzled acknowledgments of their importance would muddy the exposition of the argument without providing any offsetting benefits.

I do think, however, that the models of herd behavior and of bounded rationality are important, and I will occasionally refer to them in this book.

PART TWO

Legal Applications

4

Gifts and Gratuitous Promises

Gift-giving is the most important way in which people convert cash into reputation. By cash, I mean all those assets that either are cash or can be liquidated relatively easily. One's house, clothes, and bank account are all cash assets in this sense. Reputation refers to people's beliefs about one's discount rate and other cooperation-relevant characteristics. Most people want both cash assets and reputation: the goods are complements. If I have a terrific reputation but no cash assets, I will want to trade in some reputation for cash, so that I can buy food, shelter, and the other things I need to survive. If I have a lot of cash but no reputation, I will want to trade in some cash for reputation, because without a reputation it is hard to enter the cooperative relationships that produce valuable nonmarket goods, like security, intimacy, and political influence. People make different tradeoffs, but the extremes—embodied in the miser, on the one hand, and the saint, on the other hand—are rare.

Scholars, especially hard-nosed economists and law professors, habitually underestimate the importance of gift-giving in daily life. In fact, people spend substantial amounts of time and money thinking about, shopping for, buying, and exchanging gifts. Says a noted anthropologist of modern life: "One young mother I know laments frequently that a disproportionate amount of her discretionary income is spent on gifts. Her three children collectively attend approximately 40 birthday parties a year. Her office staff numbers four and she gives each a Christmas gift. The larger office takes collections (almost daily, she says), for flowers for employees who have babies, are ill, have experienced a death in the family, or are promoted. Her husband's family is huge, and . . . Her children's teachers receive . . . And she and her husband are frequently entertained, requiring a series of hostess gifts" (Post 1992, p. 561). Yet she cannot refuse to give these gifts without damaging important social and business relationships. Each gift is an investment in the preservation and enhancement of a favorable reputation.

People who have built up reputations over the years frequently cash them in as old age overtakes them. As people age and their time horizon narrows, they derive diminishing returns from even sterling reputations. At some point, they will want to cash in their reputations. Aging celebrities do this by making commercials. A three-way transaction occurs. The celebrity gains cash, but loses his reputation for integrity. The firm advertised loses cash, but benefits from the enhancement of *its* reputation. The bridge is the public, which learns that the celebrity has a high discount rate, and thus has little value as a cooperative partner. This is the source of our contempt for the person who sells out. The observer also learns that the firm has confidence in the value of its products; otherwise, it could not afford to throw money at the celebrity (Klein and Leffler 1981).[1] This is the source of the firm's gain. Conversely, younger people who have accumulated large cash reserves derive diminishing returns from cash and seek to convert marginal dollars into reputation. The demand for a means to convert cash into reputation conjures up supply. The suppliers are charities, which are firms whose business is to sell reputations to those who will pay for them. These entities receive donations of billions of dollars every year (Clotfelter 1992).

Gifts as Signals

The altruism of gift-giving is such an integral part of the ideology of everyday life that people who have not thought deeply about gift-giving have a great deal of trouble accepting the possibility that altruism plays only a limited role. The ideological glare of gift-giving has blinded law professors to its social function. Yet scholars outside of law see gift-giving differently. That almost all gift-giving takes place within relationships of exchange—whether overtly commercial, or implicitly within social or intimate relations—has long been understood by anthropologists and sociologists, such as Mauss (1990), Blau (1964), and Miller (1993), and historians, such as Nissenbaum (1997).[2] This chapter examines this idea using the signaling model developed in Chapter 2.

Gifts are signals (Camerer 1988).[3] They are easily, indeed unavoidably, observable and costly actions that waste resources. Gifts are costly in two ways: one must pay money or expend effort in order to acquire goods to convey as gifts; and one must invest effort or money in order to determine the tastes of the donee. A gift of cubic zirconium jewelry is less successful than a gift of diamond jewelry, because cubic zirconium is cheap. A gift of wine is inappropriate when the donee is a teetotaler. Gifts of flowers and candy are adequate but not good gifts: because almost everyone likes flowers and candy, the gift does not reveal that the donor has invested in learning the donee's tastes; however, because these gifts can be expensive and most people do like them, they can still be meaningful (as opposed to a gift of paper clips). Gifts are common signals because, uniquely among other kinds of signals, the cost of a gift is within the control of the donor, so a donor can invest just as much in a gift as is necessary to distinguish his type, and no more. By contrast, other signals—like education and discrimination—may be "lumpy," because their cost depends on factors outside the agent's control.

In order for gifts to serve as a signal, they must reveal that a person has a low discount rate. The problem is that an identical gift may be more costly for some people than for others. One important source of ambiguity is altruism. If a person gives a gift out of altruism, the gift is "cheap" in the sense that, because his utility increases as the donee's utility increases, he obtains a return on the gift that a non-altruist does not receive. Suppose that A and B have identical tastes and endowments except that A cares about C, whereas B perceives C as a good business partner. If B would obtain a discounted return of $100 from a relationship with C, then B would be willing to pay up to $100 for a gift to C that would ensure that C would realize that B is a good type and enter a relationship with B. A does not want a relationship with C, but does care about C. So A might be willing to give C a $100 gift just because A's increase in utility, resulting from C's increase in utility, exceeds the loss of utility from spending $100 that could have been used for some other opportunity. If C does not already have clues about whether A or B is altruistic, and whether A or B is a good or bad type, then the $100 gift will not enable C to determine the truth.

Gifts become more expensive as the amount at stake in a potential relationship rises. Gifts serve as signals only if they cost the donor more than the payoff from cheating in an early round; so as the amount at stake rises, this payoff rises, and therefore the amount invested in gift-giving rises. Thus do diplomats and businessmen give each other more expensive gifts than do ordinary people when they seek to enter or enhance relationships with their friends and associates. Because the value of relationships, especially business relationships, depends on the skills and endowments of people involved, wealthier and more skillful people give each other more expensive gifts than poor people do. If capital markets functioned well, poor people with excellent skills and discount rates would be seen borrowing money in order to give expensive gifts to businesses or individuals that would hire them only if they trusted them. But capital markets do not function well. Indeed, if information were sufficiently available to allow such poor people to borrow money to make gifts, then signaling would not be necessary. In fact, poor people with excellent skills and discount rates cannot borrow much money, but must signal their type by spending a lot of time working their way up from the bottom.

A second source of ambiguity in gift-giving, then, results from uncertainty about whether the expensiveness of a gift reflects the donor's discount rate, or other skills that the donor has, or the donor's possibly mistaken expectations about the skills and endowments of the donee. For example, a boss' expensive gift to the secretary might reflect his possibly false belief that she has skills that would enable her to take a management position and contribute more value to the firm than she does as a secretary. Or it might reflect simply his desire to show her that he can be trusted in the continuing boss-secretary relationship but that he has an unusually low discount rate.

We should say more about the difference between one-sided gift-giving and reciprocal gift-giving. One-sided gift-giving occurs when only one person has private information about the recipient's discount rate. For example, when individuals enter clubs and other institutions where members are relatively well-known, the entrant must engage in most of the signaling, and he may do so by holding parties, making contributions to charities, and so on. Reciprocal gift-giving occurs when both sides have private information about their type. Cash gifts can serve as signals in one-sided gift-giving. If the cash amount exceeds the payoff from cheating in an early round, the donor could not enter the relationship in order to cheat the donee. In effect, cash gifts, by offsetting any early returns from the relationship, defer compensation until a later period, where only a person with a low discount rate would enter the relationship given deferred compensation. By contrast, when both sides must signal their type, cash cannot serve as a signal, because the cash gift from the one side compensates the other side for its cash gift to the first side. The exchange of gifts is a wash, or, if the gifts are unequal, they result in a one-sided transfer. Reciprocal gift-giving requires the destruction of value.

This result explains why so much ritual gift-giving involves the exchange of gifts consisting of goods or services rather than cash. At Christmas and similar holidays, across birthdays, at dinner parties and other special occasions, cash gifts are rarely exchanged among people who belong to the same social or business circle. When foreigners stay at people's houses, they bring as gifts goods characteristic of their homeland, not cash. Jokes about revolving fruitcakes and ugly ties reflect our appreciation that much of gift-giving has nothing to do with giving people something they need. We feel compelled to give gifts even when we do not know what gifts will please the recipient and fear offending him. But we must try, in order to show that we remain committed to a relationship. If, by contrast, gift-giving were always the result of altruism, one-sided gifts would always consist of cash, because cash is almost always worth more to the donee than the goods it is used to purchase. Reciprocal gifts would not occur, as both sides would do better by not giving gifts, than by exchanging gifts that each donee values less than the cash equivalent.

Another reason for the rarity of cash gifts is that the investment of time and effort in determining the donee's tastes may be a more reliable signal than cash. A man who has a joint bank account with his wife harms her when he buys a gift for her using money that she values more than the gift. (Imagine that she earned all or most of the money in the account.) By contrast, if he uses time to determine her tastes—time that would otherwise be devoted to his own, private leisure activities—then he incurs costs without at the same time injuring his wife. The best gifts surprise the recipient but at the same time please him and seem particularly suitable for him. As noted above, flowers and candy are hackneyed gifts, because everyone likes flowers and candy and no one is surprised to receive them. In contrast, clothes that look especially good on the recipient, art that appeals to his tastes, a book that interests him but that he has not read—all of these can be good gifts. A good gift shows that the donor has developed expertise about the donee's tastes, but only someone who intends to maintain a long-term relationship would invest in developing such an expertise, because the relationship must endure for a long time before he will recover the costs of that investment. So a good gift—one that reveals that the donor has a deep understanding of the donee—is a reliable signal that the donor is a cooperator. And money could never be such a gift, since a gift of money requires no knowledge about the donee's tastes and personality (Camerer 1988, pp. S193–94).

Once two cooperators match up, they form a relationship in which they produce and divide a cooperative surplus—one that could not usually have been specified in advance and enforced through a legally enforceable contract. Pursuant to this trust relationship, each party takes any action that produces more benefits to the other party than costs for the first, or at least any such action that the other party can observe and evaluate. Transfers and promises will often look gratuitous in the sense that each such action is not specifically determined in advance as part of a quid pro quo but is whatever action the actor believes will maximize the surplus. The apparent gratuitousness of trust-related behavior, however, should not conceal the fact that the parties enter the relationship for the purpose of obtaining personal (non-altruistic) economic gains.

For example, a university might give a prospective employee dinners, trips, meetings, flattering phone calls, and so on, which are not only costly to the university but of minimal value to the prospective employee. The high cost of these gifts signals that the university seeks a long-term relationship. Their low value deters prospective hires with no interest in academia from feigning an interest in academia in order to receive some attractive but one-time gifts. The academic's willingness to accept the gifts signals his interest in a long-term relationship. Universities offer much less valuable gifts—or none at all—to prospective lecturers, adjuncts, and similar short-term employees, because no long-term gains justify the costs. Once a good academic and university form a relationship, the academic engages in value-maximizing activity on behalf of the university (writes papers and serves on committees, even though he has tenure) and the university does the same on behalf of the academic (gives him regular raises and so on, even though he has sunk roots into the community and cannot easily leave). Notice that when the academic serves on a committee, this action is gift-like—a transfer that does not explicitly call for a reciprocal transfer—but it is really taken pursuant to a contract-like relationship that cannot be enforced legally because of information costs.

It is thus important to distinguish between gifts that are made for the purpose of signaling and gifts that occur as part of the loose quid pro quo in a trust relationship, which are not really gifts at all. The signaling gifts correspond to "formal" gifts discussed previously—gifts that are ritualistic and not jointly value-maximizing, such as an exchange of fruitcakes at Christmas. These gifts are motivated by the desire to enter or continue a relationship of trust. The other sorts of gifts—call them oxymoronically but appropriately "exchange gifts"—are best understood as any transfer from one party in a trust relation to the other, which benefits the recipient by more than it costs the donor. These gifts are motivated by a desire to obey the terms of the relationship so as to continue to benefit from it. Exchange gifts benefit the donee more than they cost the donor; signaling gifts frequently cost the donor more than they benefit the donee. Exchange gifts are transferred as the opportunity arises; signaling gifts are transferred in a ritualized fashion.

Why is gift-giving ritualized? The answer is that because people fear that their gifts will be misinterpreted, they imitate gift-giving behaviors that over time have become patterned and well-understood. The boss who wants to congratulate his secretary for good work should not give her a gift on Valentine's Day. Valentine's Day is a focal point for romantic gift-giving, not business gift-giving. Secretaries' Day and Christmas are focal points for giving gifts to secretaries. Indeed, holiday gift-giving generally is focal: people understand that holidays are times for people to enter or enhance relationships through reciprocal gift-giving. I discuss focal points in greater detail in Chapter 7; for now, it is important to understand that characteristics of gift-giving rituals will often be a function of historical accident.

I have emphasized the importance of trust relationships in business, where they allow parties to exploit surpluses unobtainable through contractual mechanisms because of the cost of information. They are also important in family relations, where legal norms as well as information costs restrict the ability of family members to make and enforce marriage contracts pertaining to household production. The most vivid example of their importance, however, comes from the political arena. Law and public policy bar political bribery, so lobbyists give politicians gifts in the hope of influencing their votes. The initial gift is a signal from the lobbyist that future gifts will be forthcoming if the politician acts properly. Later gifts are rewards for earlier behavior and promises of more to come.[4]

One might argue that public charitable gifts are actually disguised sales. The university sells its right to name a building after someone else to the "donor" in return for the latter's money. This description, however, does not capture what is special about the transaction. Universities still resist the temptation to sell the right to name buildings to wealthy scoundrels who seek a cheap way to repair their reputations. Like the politician who receives a campaign contribution, the university maintains a formal, even if rarely exercised, right of discretion. The reason that universities do not simply sell titles and positions is that their prestigiousness would be destroyed if these were routinely sold, just as the prestigiousness of the Nobel prize would be destroyed if it were auctioned to the highest bidder. The Reformation taught the Catholic Church a similar lesson about the difference between the price and the value of indulgences. In a phrase, people value reputations for generosity, ingenuity, and fair-mindedness, but if one could purchase such reputations, then they would cease to exist.[5]

Alternative Motives for Giving Gifts

Altruism. As previously noted, people give gifts out of altruism as well as out of a desire to enter or enhance relations of trust. There is a widespread intuition that altruism is the *sole* motive for giving gifts.[6] According to this view, a donor gives a gift to a donee when the donee's increase in utility causes an increase in the donor's utility that exceeds the decline in the donor's utility resulting from the expenditure of time or money on the gift.[7] This view is wrong.

Consider the donor who gives gifts to a charity because he cares about the well-being of the charity's beneficiaries. Assume that many people share the donor's concern about the well-being of the beneficiaries and make contributions, and that each person takes every other person's contribution as given. Then any donor would decrease his contribution to a charity when other people increase theirs. Because the well-being of the beneficiaries increases through the other donations, the donor in question has a stronger incentive to spend his money on something else. This result conflicts both with intuition and with evidence (Sugden 1982). In addition, the altruism theory cannot explain why charitable gifts are almost always distinguished by the high degree of publicity that they receive, and it cannot explain other patterns of charitable gift-giving.[8]

As mentioned above, altruism also cannot explain reciprocal gift-giving. If X and Y exchange gifts of equal value at Christmas, then each ends up with a good rather than money that is worth no less and probably more to him than is the good. (In the absence of signaling value, a good cannot be superior to its money equivalent, because the donee can always buy that good if he prefers it to anything else, and he can use the money for other things if he does not.) If the parties entered an agreement not to exchange gifts, then each party would gain from (1) *qua* egoist, the fact that he has cash rather than a good of equal or less value, and (2) *qua* altruist, the fact that the other party has cash rather than a good of equal or less value. But such agreements are rarely observed.[9]

Status Enhancement. Some scholars argue that people donate large sums of money in a public way in order to enhance their status. To analyze this phenomenon, many scholars treat status as an exogenous factor. They assume that people have a "taste" for status, which means that they obtain utility when they have more of some status good than others do (see, for example, Glazer and Konrad 1996, Bernheim 1994, Coleman 1990, pp. 129–31, Frank 1985, McAdams 1992, Hirsch 1976, Scitovsky 1992).[10] As I mentioned in Chapter 3, I do not think it is necessary to make such an assumption in order to explain status-seeking behavior.

Define status as the attractiveness of a person as a cooperative partner. People are drawn to high-status people, because they would obtain more from a cooperative relationship with those people than with lower-status people. Status, then, is a function of people's beliefs about the characteristics that a person has, characteristics including discount rate of course, but also skills such as athletic ability or business acumen, endowments, and features such as beauty or charm or generosity. A consequence of this definition is that a person with certain skills, endowments, and features might be high-status in one group and low-status in another group. For example, a movie actor has high status among movie fans, which includes most ordinary people. A genius mathematician might have high status among academics. A local politician has high status among his constituents but not among foreigners who are unaffected by him, or among religious initiates who have withdrawn from the world. A great baseball player but terrible companion might have high status among baseball fans but low status among his teammates or neighbors.

The concept of status draws a sharp line between a person's unobservable characteristics and the beliefs that other people have about those characteristics, which are based on the first person's observable actions and qualities. Everyone has a powerful incentive to change those beliefs in a way that increases his status even if not in a way that increases their accuracy. Indeed, people want to influence the beliefs of whole populations, as a way of "advertising" for potential cooperative partners.

Assume that when people look for cooperative partners, they understand that the information revealed by gift-giving is quite noisy, so rather than deduce a range of discount rates (and other characteristics) from a gift, they instead rank people according to the value and extent of their gift-giving activity. Now people will always have beliefs about others based on past actions that are unrelated to gift-giving, some of which may be based on their performance in cooperative relationships, and some of which is based on other forms of signaling. For example, a tycoon may have status because he has an important position in a well-known firm. This suggests that he has significant skills and that he can be trusted (he has a low discount rate). The tycoon also lives in an expensive house, sends his children to expensive schools, wears expensive clothes, and takes expensive vacations. These instances of conspicuous consumption cause people to believe that the tycoon is indeed wealthy, and thus that he might have the standard wealth-generating characteristics. Now, if in addition the tycoon makes generous gifts to charitable organizations, people may increase their estimates of his wealth and believe that he is warm-hearted and generous as well. The enhanced reputation for wealth, skill, and so on will attract people to the tycoon: the tycoon will find it even easier to form social and business relationships. This is why gift-giving is such an important part of the effort of the nouveaux riches to break into "society."[11]

If enough people care sufficiently about their reputations (for being generous or for being wealthy), almost everyone will conform to a norm of gift-giving behavior; almost everyone will give more wealth than he would if he did not care about his reputation; and people's charitable contributions will not be as sensitive to the well-being of the donee as the altruism theory predicts. These arguments will be discussed in greater detail subsequently.

I should make three further points. First, a donor may seek to signal his wealth or generosity to some people but not other people. A person may give an anonymous donation in order to signal these qualities to a spouse. A person may give a donation to a charity in order to signal his wealth or generosity to friends and associates without caring about the opinions of the actual recipients of the gift. A tycoon donates money to the opera rather than to movie theaters because he cares about the opinions of the elites who attend the opera, not the masses who go to movies.

Second, I should emphasize that an individual *donee* suffers a loss of status. (Interestingly, an institutional donee does not suffer a loss of status; nor do its directors or officers.) The acceptance of an expensive gift, when no reciprocation is expected or occurs, signals that the donee does not have much wealth. If status is a function of perceived wealth, the donee will suffer a loss of status. This is why a person who wants to humiliate another person can do so by offering him charity. In order to avoid losing status, even donees in great need will sometimes refuse to accept charity, or will insist that the charity be hidden, or will attempt to reciprocate after receiving charity—all of these are favorite subjects of Victorian novelists.

Third, status-signaling, which is really just a form of competitive trust-signaling, must be credible. Since charities understand that people give them gifts in order to signal their status to others, charities take pains to publicize the names of donors. Your colleagues at work may never really believe your stories about your lavish vacations or about the value of your art collection, but they cannot question your entitlement to the label "Patron" when they see your name under that category in the program at the opera (Glazer and Konrad 1996, pp. 1019–20). As I will discuss below, one theory of charities holds that they exist in part to enable wealthy people to send credible signals of their abilities and endowments to the limited populations from which they seek cooperative partners.

The Social (Dis)utility of Gifts

What does gift-giving have to do with social norms? One common argument is that social norms prohibit undesirable gift-giving. When a person gives too lavishly or too meanly, we disapprove of him and may even sanction him because he violates social norms. I argue instead that "social norm" is just the label we attach to equilibrium behavior. Social norms do not "cause" people to do anything, except insofar as they try to figure out what is optimal by imitating others, and we cannot determine whether social norms are good or bad without understanding the incentives for the underlying behavior. In a community in which people give lavishly because of a destructive status competition, people give lavishly in order to reveal their type. People who do not give enough are identified as bad types and shunned. It is fine to say that the lavish donors conform to the social norm, while the mean donors violate the norm, as long as the rational basis of the behavior is understood. People in a different community might view this behavior from afar, and ridicule the lavish donors while sympathizing with the mean donors. In this other community, lavish gift-giving is not an appropriate signal. But the existence of this community is no comfort to the mean donors in the first community, except that if they lose the status competition because of their relative poverty, they may emigrate from the first community and into the second.

Waldfogel (1993) argues that in 1992 Christmas-giving produced a deadweight loss of between 4 and 13 billion dollars. The figures are generalized from a survey in which respondents attributed to gifts they had recently received a value that was lower than the purchase price of the gifts. One interpretation of this result is that people prefer money to goods, because money is fungible and can be used to buy whatever one wants the most. Thus, whenever a person gives goods (or services) as gifts, this produces a deadweight loss equal to the difference between the value the donee attaches to the goods and the value the donee would attach to the amount of money used to buy the goods. When this behavior occurs on a large scale, such as at Christmas, billions of dollars are lost.

Waldfogel's results raise the possibility that gift-giving generates enormous social costs. Is this at all plausible? The answer is surprisingly complex. Even straightforward altruistic gifts, which involve no signaling at all, can be socially costly. Although this is to the side of our main interest in how signaling generates social norms, a discussion of these costs is worth a brief digression.

It might seem obvious that altruistic gifts have social value: gifts make the donor and donee better off without making third parties worse off. They make the donee better off because the donee prefers the gift to nothing, and they make the donor better off because the donor derives utility from the donee's increase in utility.[12]

There are several problems with this argument. One problem is that the donor takes account of the donee's utility only as it is translated into his (the donor's) own utility function, and does not account for it separately as it would be in a social welfare function (Kaplow 1995, Friedman 1988). For example, suppose that the donor proposes to give the donee a painting worth \$100 to the donor and \$200 to the donee, and that the donor values the donee's well-being at a discount of 40 percent. The donor makes the gift because he gains \$120 worth of utility while losing \$100. In addition, the donee gains \$200 from the gift. The gift, which would result in a Pareto improvement, will occur. Now suppose that the donor values the donee's well-being at a discount of 60 percent. The donor does not make the gift because its cost (\$100) exceeds the benefit to him (\$80). However, if the donor did give the gift, the donee would gain \$200, which exceeds the donee's loss. The gift will not occur, but if it did, it would increase total utility, though it would not result in a Pareto improvement. Although it is theoretically possible that the donee would offer to pay the donor, say, \$21 to make the gift, thus making both the donor and the donee better off and enabling the "gift" to occur, this Coasean bargain rarely, if ever, happens. A possible reason for this bargaining failure is that recipients are never quite sure whether a gift is given for altruistic or for trust-enhancing reasons. Attempting to bargain over a trust-enhancing gift is terribly improper, as it suggests that the donee is neither a cooperator who seeks a relationship nor a cooperator who does not seek this particular relationship, but rather an opportunist seeking to get a signaling gift at no cost to herself—something that would be in no one's interest to admit. Bargaining is appropriate when the donor is an altruist; but a pure altruist would give money, not goods, so the bargaining would not be necessary.[13]

Altruism produces other problems as well—it can lead to overconsumption by donees who seek to provoke altruistic donors to give more than they want to give, and it can interfere with cooperation by undermining the credibility of the altruist's threat to retaliate against someone who cheats (Bernheim and Stark 1988).[14] But let us turn back to gifts as signals.

At first sight, one might conclude that the use of gifts to enter and maintain trust relationships is a socially valuable practice. A problem, however, arises from the struggle between bad types, who attempt to mimic good types, and good types, who attempt to distinguish themselves from bad types. Suppose, for example, that everyone—bad and good—conveys signaling gifts. If the bad types would lose more by being identified, and thus avoided, than they would gain by saving on the cost of gift-giving, then they will not deviate from the equilibrium; but neither would the good types, for then they would be mistaken as bad types and lose the gains from the trust relationship. Both types would be better off if they could commit themselves never to give gifts, since they would then save on the costs of gift-giving without losing any other benefit. But there is no mechanism by which such a commitment could occur. Now, suppose that only good types give gifts and bad types do not give gifts. It may nevertheless be the case that all good types would be better off if none of them gave gifts (the cost of giving gifts exceeds the gains from cooperative relationships when bad types and good types cannot be distinguished), while at the same time no single good type would deviate from the equilibrium (the cost of giving gifts is less than the cost of being mistaken for a bad type and avoided). Both examples show that one cannot say in the abstract whether trust-enhancing gift-giving equilibriums are efficient.[15]

The problem is more serious when people compete for cooperative partners in the public arena. When a tycoon endows a university building or chair in his name, clearly the university benefits.[16] Students, professors, and others gain to the extent they share in the consumption of a public good supported by the donations. However, people also lose to the extent that they care about their own status, and to the extent that status is a function of the relative size of one's contributions. When a large donation is made, status-conscious observers must dig into their pockets just to maintain their rank. But if they do so, then the original donor does not obtain any status despite the costs he has incurred in conveying the gift. If enough people seek to enhance their own status by giving gifts, then the cost of any particular gift necessary to enhance the status of the donor must rise.

The competition for status is a prisoner's dilemma. Each potential donor faces a choice between making a gift, in order to assert his status, and not making a gift. If he expects other people to give gifts, he should give a gift, because otherwise he will lose status. If he expects the others not to give gifts, he should give a gift, because by doing so he can gain status at the expense of the others. Therefore, each potential donor will give the gift. But because everyone gives a gift, no one gains status (because giving a gift does not distinguish one from the others); and this outcome is inferior to the case where no one gives a gift. In the latter outcome, no one gains status; but at least no one incurs the expense of the gift (Frank 1985). For example:

> The view was repeatedly expressed that New York philanthropy had changed during the past twenty to twenty-five years, so that one could now "buy in" to a position of prominence. One person said that there used to be a tight circle, but "what blew that apart were the [Xs] and people like that coming in who nobody knew . . . what we call nouvelle society. It did change things. All of a sudden these people started giving huge amounts of money. You had to notice. *But it also made people look bad who were moderate donors and they didn't like losing their position.* (Ostrower 1995, p. 43; brackets and ellipsis in original, emphasis added)[17]

Some might argue that this outcome is socially beneficial, because the status competition leads to the provision of public goods (McAdams 1992). The external value of donors' contributions to the public good exceeds the cost to themselves. The problem with this argument is that there is no reason to believe that the equilibrium at which the donors give gifts matches the socially beneficial level of gift-giving. It might be lower or higher.

In addition, the equilibrium will favor some charities at the expense of others in a way that bears no relation to their social utility. The amount of money raised by a charity from status seekers is a function not of its social value, but of the extent to which its purpose justifies expenditures on media that widely disseminate its list of donors.[18] Status-enhancement is most likely to occur when the donor's generosity can be prominently memorialized: when a building or chair is named after him, or when names are listed on plaques or in programs that are widely seen. Many charities, such as anti-poverty charities, cannot offer these services or can do so only by distorting their mission (constructing buildings, for example, rather than providing services), or simply by providing donors with a means of publicity. Hence the incongruous sight of catered black-tie balls held for the purpose of raising money for poverty relief. Because there is no relation between the public prominence of a charity, or the prominence of its activities, or the ease with which its mission justifies the production of buildings or permanent institutions, and its social value, status-driven philanthropy cannot be expected to maximize social welfare.[19]

Evidence for the social disutility of status-enhancing gift-giving may be found in a wide range of contexts. In many societies, especially highly stratified ones, people sometimes ruin themselves in an effort to meet gift-giving obligations. The dowry system in India, for example, leads to far too much mayhem to have a plausible claim as an efficient institution (Roulet 1996, Teja 1993). Missionaries attempt to abolish potlatches, not just because potlatches reflect non-Christian beliefs, but because they appear to waste valuable resources. "[M]ost of the Indians are industrious; if they were less wasteful and could be made to abolish the potlatch system among themselves, I believe there would be very little real need," wrote one missionary of a group of Alaskan Indians (Simeone 1995, p. 26).[20] The dysfunctional nature of some forms of gift-giving can be seen in the occasional legal responses to them. India, for example, has laws against dowry; and, to take a minor but telling example, when the practice of tipping came to the United States in the beginning of the twentieth century, many states prohibited it by law (Zelizer 1994, pp. 94–99). The behavioral regularities that emerge as a result of using gifts to send signals are described in common speech as social norms. So our broader lesson is that social norms can be undesirable, and legal intervention to change or weaken them may be justified.

Legal Implications

The law influences gift-giving norms in many ways. I will focus on three of interest: contract law, fraudulent conveyance law, and the law of nonprofits.

Contract law. Much has been written about contract law's mysterious treatment of gratuitous promises.[21] Courts generally do not enforce gratuitous promises except those that are directed to charities and are sufficiently formal, or that have fairly obvious business purposes, such as option contracts. At the same time, courts do not treat gratuitous transfers differently from the way that they treat commercial transfers, except that (as described in the next section) gratuitous transfers are more vulnerable to attack on the grounds of fraudulent conveyance. An adequate explanation for the law's treatment of gift-giving must show (1) why gratuitous promises are given less protection than non-gratuitous promises, while (2) gratuitous transfers are given about the same level of protection that non-gratuitous transfers receive.[22]

The previous discussion of the social utility of gift-giving does not supply easy answers. If a gift is motivated by altruism, it should not necessarily be restricted; indeed, it might be necessary to subsidize it in order to ensure that the donee's utility is given proper weight (Kaplow 1995). I have analyzed these problems elsewhere, and since the analysis is complex, and only tangentially related to social norms, I will not repeat it here (E. Posner 1997a).

What about the legal implications of gifts as signals? Parties in a trust relationship will not usually sue each other when disputes arise, relying instead on nonlegal sanctions. But the law only matters when a lawsuit occurs. One way to explain why parties that appear to be in trust relationships might sue each other is to note that two parties in a single long-term relationship are likely to obtain the largest gains by exchanging a mix of legally enforceable and legally unenforceable promises. For example, the academic can successfully sue the university for violating its promise never to fire him, but the academic can probably not successfully sue the university for failing to give him regular salary increases or for giving him onerous committee assignments. This mix of enforceable and unenforceable promises reflects the judgment of the parties that a court, with its superior sanctions but inferior information, could do an adequate job of identifying extreme cases of opportunism but not minor cases of opportunism.

When the parties sue each other, the value-maximizing court must distinguish between intendedly legal and nonlegal promises. To be sure, parties can facilitate this judgment by stating clearly in the contract which promises are meant to be legally enforceable and disclaiming those that are not. The costs of anticipating contingencies and accounting for them, however, prevent the parties from describing all enforceable and non-enforceable promises in the contract. Accordingly, a default rule is necessary to determine the enforceability of promises about whose enforceability the contract is silent.

One candidate for such a default rule is the consideration doctrine. Although often understood today to mean that gratuitous promises are not enforceable, the consideration doctrine historically had a narrower meaning. Under the consideration doctrine, a promise was enforceable only if motivated by a *specific,* or bargained for, performance or return promise. A promise might be unenforceable under the consideration doctrine just because it was a gift promise; but it also could be unenforceable because the return performance or promise could not be clearly identified. Thus did the consideration doctrine forbid the enforcement not only of gift promises, but also of requirements and output contracts, firm offers, contract modifications, and other promises made in return for something that was vague. Now courts might resist enforcing such promises because of the difficulty of proving them. Probably they resisted enforcement in the past because they feared that if they did enforce such promises, it would be too easy for people to make a fraudulent claim in front of the court that someone had made a promise that never in fact occurred. (In this respect, the consideration doctrine is a contract formality like the Statute of Frauds.) But it is also possible that courts understood that for complex and indefinite contracts, parties could deter opportunism through nonlegal mechanisms more effectively than the courts could through legal mechanisms. Doctrines that prevent judicial enforcement are justified on the grounds that judicial enforcement would interfere with trust relationships.

The idea with which we started this chapter—that a gift sets up, or is part of, an indefinite exchange—converges with the idea that the consideration doctrine prohibits the enforcement of indefinite exchanges. The consideration doctrine prohibits the enforcement of gift promises not because of a policy against gift-giving but because courts want to encourage parties to be specific about the content of their exchanges in order to ease the judicial burden of interpretation. Similarly, in the past courts resisted enforcing firm offers and requirements contracts not because they were socially undesirable, but because, like gift exchanges, they were vague. But judicial convenience had to give way to commercial exigency. Courts gradually realized—or, at least, came to believe—that parties would prefer the uncertainty of judicial enforcement of vague terms to the certainty of non-enforcement, and over time courts yielded to entreaties to enforce vague contracts. Whether courts will treat gift promises in a similar way remains to be seen. They should only if the social value of enforcement of such promises exceeds the cost of fraud—an empirical question.[23] Our analysis suggests that gift promises are different from commercial transactions generally, and their social value is more ambiguous. So further liberalization of the consideration doctrine should occur only with circumspection.

Fraudulent conveyances. Stylized facts, drawn from a recent case,[24] raise an interesting question about fraudulent conveyance law. Donor runs a Ponzi scheme from which he earns $20,000. He spends $10,000 on various goods and services and he donates another $10,000 to a church of which he is not a member. The issue is whether the donation to the church is a fraudulent conveyance.

Fraudulent conveyance law makes a distinction between conveyances for which the debtor receives a fair consideration and conveyances for which he receives little or nothing. If the debtor buys a car for $20,000 without any intention of defrauding his creditors, the sale is not reversed under fraudulent conveyance law, although the car can be recovered by the creditors. If he gives $20,000 to his best friend without any intention to defraud his creditors, then the gift is reversed. The purpose of fraudulent conveyance law is to prevent debtors from hiding assets with friends or relatives, then obtaining them after creditors have given up the chase.[25] This purpose follows from the plausible theory that most debtors would give up their ability to hide assets in anticipation of default in return for a lower interest rate (Baird and Jackson 1985).

Fraudulent conveyance law thus assumes that the average donor (the debtor) and donee have a trust relationship. If they did not, the donor could not expect to recover the goods after the creditors gave up the chase. But in our hypothetical case the donor and the church do not have a trust relationship. The motivation of the donor was altruism or, possibly, status enhancement. In either case, the donor will not be able to obtain his funds from the church after he gets out of prison. The policy of fraudulent conveyance law does not justify revocation of the gift.[26]

One might respond that the policy of fraudulent conveyance law is, on the contrary, simply to maximize the value of the debtor's estate. But if the debtor purchased and consumed a plane ticket with the $10,000, the airline does not have to repay the creditors. The reason is perhaps that it is cheaper for the creditors to screen against fraud or, more specifically, take precautions to ensure that they do not lend to people who are likely to commit fraud, than it is for airlines to screen against payments out of funds tainted by fraud. But if this is true about airlines, it is also true about charitable organizations.

Nonprofits. I started this chapter by observing that cash and reputation are complements. Cash enables a person to buy goods and services, but it does not enable people to "purchase" relationships. Reputation enables people to enter personal relationships and obtain nonmarket goods, like political offices, that cash cannot buy. If one has a lot of cash but little reputation, one will want to transform cash into reputation. How might one do this?

The way to transform cash into reputation is to give it away, in the right way. Donations to charity show people that one is wealthy and generous, and that one has a low discount rate. Bad types do not want friends as much as good types do, because friendship requires immediate and significant investment in return for an uncertain long-term gain. So giving away money in return for nothing is a way of distinguishing oneself from the bad type. But it is important that people be able to observe one when one makes contributions.

This is harder than it might seem. Writing checks in front of strangers is time-consuming, and anyway hardly credible, since one can tear up the check after writing it. People have better things to do than stand around watching others distribute cash. What are needed are institutions that will certify the transfer of money from donor to donee.

These institutions exist. We call them operas, symphony orchestras, environmental organizations, and universities (Glazer and Konrad 1996). One of the purposes of these institutions is to take people's money in a public way. They take your money, and they either directly advertise the gifts (in programs and on plaques) or they provide the donor with the means to self-advertise (like stickers or decals). They "sell" reputation; the price is a cash "gift."

One might object that this theory does not explain why charitable institutions produce operas, feed the poor, and educate the young. Why don't institutions arise that simply certify in a public way that certain people have given them a large sum of money?

The answer is that these institutions would have trouble attracting people's attention. The closest analogies are magazines like Forbes, which list the wealthiest people in the country. But these magazines must attract readers by providing interesting articles, and, because they have limited space and face readership constraints of various sorts, they can publicize only the extraordinarily rich—as opposed to the richest people in a city or town or community.

A better analogy to charitable organizations is television. Advertisers give money to television stations not because they care about the shows, but because they want the public to see their advertisements. But the public will see their advertisements only if they watch television, and they will watch television only if they enjoy the shows, so the stations must invest in good shows. Similarly, donors do not necessarily care about the opera, but give money because they want other operagoers to see their names in the program. The public will see these names only if it enjoys the opera. So institutions have an incentive to produce something (like opera) that people like.

A donation to a charitable institution is like an investment of money. In return for the cash "investment," the donor has a (residual) "right" to recognition over time just as an ordinary shareholder has a right to dividends over time. The donor's right, unlike the shareholder's right, cannot be legally enforced. The donor cannot sue if the donee produces a bad opera that no one sees, resulting in no recognition to the donor. Institutions thus can attract donations only by establishing a record of successful publication of the names of donors. The opera company has an incentive to produce a good opera because a failure to attract audiences will prevent it from raising more money in the future.

Why are operas non-profit institutions? Hansmann (1996) notes that the managers of operas and similar institutions are better monitored by consumers than by outside investors. This point is plausible, but it does not explain why operas are not consumer-owned cooperatives. Hansmann's explanation for this fact is that the non-profit form allows operas to engage in price discrimination. If operas made profits, people would not donate money to them, because their money would go into the pockets of shareholders. If donors like the shareholders so much, better to give them cash gifts directly! People donate to operas in the hope that their donations will improve the quality of the operatic productions. But this claim assumes that people are irrational. They ought to free-ride. The implausibility of this claim can be seen in Hansmann's similar argument about why universities take donations rather than charge higher prices for education. He argues that students enter an "implicit commitment" to donate in return for a below-cost education (Hansmann 1996, p. 233), but does not explain how this commitment is enforced. And if it is not enforced, then we are left with the question of why graduates give their alma mater money for nothing.

An alternative theory for the non-profit status of operas, universities, and related institutions is that donors earn returns not from having the opportunity to watch an enjoyable performance or seeing young people receive educations, but from experiencing an enhancement in reputation—an enhancement that takes the form of greater recognition, invitations to high-society dinners, new business contacts, and so on. This reputational residual takes the place of the cash residual that the shareholder receives from for-profit corporations. If an opera or university does poorly, the donor will fail to experience the gain in reputation that he seeks, and shift funds to an institution that can produce a better return on his investment. This non-market discipline substitutes for the market discipline that keeps for-profit institutions in line.

Nothing prevents an opera from shifting from non-profit status to for-profit status. Hospitals have been undergoing this change for many years. The choice between non-profit and for-profit status depends on the demand for reputation, on the one hand, and the efficiency with which the non-profit can supply it, which is itself a function of the relative effectiveness of market discipline, and the institution's ability to publicize the names of donors in a way that reaches an elite audience. What is distinctive about the opera, and what places it in the same category as the university, is that the audience of the donation comes from the elite.

This analysis raises many normative questions. It implies that because monetary returns are taxed and reputational returns are not taxed (if they result in non-market goods like friendship or political influence), people will overinvest in reputation. This raises the question why tax breaks for non-profits and for charitable donations are necessary. But this question and others are best left for future research.

5

Family Law and Social Norms

Any useful theory about law and social norms cannot ignore the family. The traditional family is the cleanest example of an institution that maintains itself through nonlegal sanctions rather than through legal enforcement, and it gives us a clear view of the opportunities and dangers that such institutions present to governments. On the one hand, people who form family relationships obtain valuable goods and services that are for the most part not available from the market and other outside institutions, and they do so without depending on formal contract law backed up by the threat of government enforcement. On the other hand, to the extent that people do not rely on government enforcement of marital obligations, they become highly vulnerable to opportunism within the family. Although generally speaking governments have given families a great deal of autonomy, the level of intervention has varied considerably over time. In the United States, family autonomy has declined over the past thirty years; however, as a matter of law and practice, the family remains fairly autonomous.

These observations raise a number of questions. Why do governments usually grant autonomy to families? If the answer is that families produce goods for their members more effectively than either the market or the political system could, then one must explain what characteristics of the family enable it to do so, and one must explain why the law regulates marriages in such a peculiar way. Marital law, for the most part, heavily regulates entry into marriage and exit from marriage, while declining to enforce obligations or remedy wrongs that arise within a marriage. Contract law, by contrast, requires parties entering a contract to observe few formalities. One does not need a license to enter a franchise contract in the way that one needs a license to enter a marriage. Contract law, unlike traditional marital law, does not interfere with efforts to end relationships. Finally, and most significantly, contract law will enforce whatever promises people happen to make, no matter how idiosyncratic, as long as they are legal. Traditional marital law imposes on people a standard set of obligations from which they may not deviate by agreement, though this has been changing over the last thirty years.

This chapter uses the signaling theory from Chapter 2 to explain the general structure of marital law, and why it differs in these ways from contract law. The theory relies on the premise that nonlegal sanctions play a more important role in family life than they do in ordinary commercial life—although as I argue in later chapters, this distinction is frequently exaggerated.

The Marital Surplus

People often object to formal descriptions of family life, but they are useful because they permit comparisons of behavior across different areas of social interaction. So let us start by defining a "marital surplus," which will refer to the goods and services that a person obtains by entering a long-term, exclusive, intimate relationship with a person of the opposite sex (or even the same sex, on which see below). The marital surplus consists of the benefits from having children, the emotional benefits from companionship and sexual intimacy, mutual aid, and many other valuable goods and services.[1]

To be sure, even this general statement is not very accurate. As one moves from time to time, and region to region, one finds a great variety of family arrangements. People can and do raise children by themselves, and the role of either the father or the mother in producing a child can be minimal or anonymous. The father might be an anonymous donor of sperm. The mother might be a donor of an egg, or the carrier of a fetus with which she has no genetic relation. Historically it was routine for people to enter marriage solely for the purpose of obtaining children or for economic advantages, and with no intention of forming a deep emotional relationship. And aid can be obtained from one's friends, one's birth family, and even market institutions rather than from a marriage. Still, I will assume that the cluster of benefits identified above is a goal of most people at the current time in the United States *and* that these benefits can be obtained only through a long-term relationship with a single person. Although I will call these benefits the "marital surplus," I assume that the long-term relationship necessary to produce them can be achieved either though a legally sanctioned marriage or through an unsanctioned "informal marriage" or long-term cohabitation.

Because the marital goods are collective goods, actions that maximize the marital goods will not always be in the self-interest of the spouse who should perform them. I will describe the value-maximizing actions as the marital "obligations." Marital surplus is generated only if the spouses can be deterred from shirking these obligations. The state might deter some violations, or it might not; if it does not, then the parties have to rely on themselves or on other individuals. Finally, note that because the marital surplus that can be obtained will vary among couples, so will the marital obligations that maximize that surplus. Some couples do not insist on sexual exclusivity; others do.

Parties in or entering a marital relationship face two groups of strategic problems. The first group of problems arises in what can be called the *search* or *courtship stage* when men and women choose mates from the relevant population. The second group of problems arises during the *relationship stage* when each party must cooperate in order to create the marital surplus. If the search stage yields a marriage between people with similar interests and low discount rates, cooperation in the relationship stage is likely to be successful.

One more point is about the division of labor in the supply of the marital surplus. The "traditional" family in the United States consists of a man engaging in market labor and a woman who has primary responsibility over the running of the household and the raising of the children. Thus, each partner specializes in a particular activity. In the more "modern" family the man and the woman both engage in market labor and contribute to the household, although they also may obtain more "marital" benefits from the market, such as cooking, cleaning, and childcare. As the husband and wife contribute to the supply of the marital surplus, they also consume the marital surplus, sometimes equally and sometimes unequally.

The Courtship Stage

Any reader of Jane Austen's novels can tell you that courtship was a crucial element of social life in early nineteenth century England, particularly for a young woman of elite society, whose future depended to a great extent on whether she made a good match. Much of the challenge of courtship was distinguishing the man who would make a good husband (because he is charming and sober and rich) and the man who would make a bad husband, while at the same time persuading suitors that one would make a good wife. Signaling went in both directions, of course: the man sought to distinguish the good wife and the bad wife, and to persuade or deceive women into thinking that he would make a good husband. The skillful players punctured appearances and discerned the reality underneath. The game was strategic and was played with the utmost seriousness by both sexes. Although less is at stake today, there still is a great deal at stake, and courtship remains a game played with intensity.

People send signals that will reveal whatever characteristics are demanded by the people they want to marry: healthiness, including fertility; wealth and wealth-generating assets including human capital; charm, humor, and other qualities that make for an attractive companion; and—what I will of course emphasize—trustworthiness, by which I mean the propensity not to cheat in repeated prisoner's dilemmas, or a low discount rate. Holding all the other elements the same, people prefer to match up with those who are generally trustworthy, because only trustworthy people will incur the short-term costs necessary to produce the long-term marital surplus.

During the courtship stage, then, people send a lot of signals, but mainly they signal in an effort to show that they are good types and not bad types.[2] Courtship signals are many and varied. One example is the long period during which the two people must confine their attentions to each other and ignore everybody else. This period of exclusivity is a straightforward signal that (1) a person has a low discount rate; and (2) the person is committed to the proclaimed object of desire, rather than to someone else.

Another courtship signal is gift-giving, culminating (in some times and places) in an expensive engagement ring. Gifts signal that (1) the donor is willing to incur costs—both monetary and temporal—and thus that the donor has a low discount rate; and (2) the donor is wealthy, and thus will be able to support the donee. As discussed in the prior chapter, there is some ambiguity: if a donor is sufficiently wealthy, an expensive gift does not reveal as much about his type as an identical gift from a poorer donor reveals about *his* type.

Celibacy can be a signal of one's discount rate, but so can sex. A man's lengthy and celibate courtship of a woman may be a reliable signal of his interest in a long-term relationship; however, a woman's willingness to engage in pre-marital sex can be a reliable signal of her type to the man if she thereby risks becoming pregnant and dependant. (Unprotected sex is an even more reliable signal.) Pre-marital sex is also a way for partners to show that they are healthy and able to engage in sex, and that they find each other mutually attractive. But how can the man signal his type by remaining celibate with a woman who signals her commitment by engaging in sex? Sex would have to occur after a delay but before a binding commitment.[3]

Courtship signaling culminates in the marriage vow. The vow, unlike gift-giving and celibacy, is not intrinsically costly but is a form of speech that has signaling value only because others observe the vow and will draw negative inferences about anyone who subsequently violates the vow. A vow also has signaling value if the state enforces it. Reputational harm, in the first case, and legal sanctions, in the second, are not necessarily costlier than each other, and neither is necessarily higher than the loss of time, effort, and wealth that results from the other forms of signaling. A religious ceremony might bind the participants more tightly than legal sanctions. An expensive gift might more effectively bind the donor to the recipient than a legally or socially enforceable wedding vow. The significance of the formal marital vow varies among cultures. One might hypothesize that in cultures where nonlegal mechanisms of enforcement are powerful (for example, where groups consist of relatively few people, people are dependent on each other for nonmarket goods and services, and exit is costly), formal vows of marriage are relatively unimportant. In determining the optimal marriage law, one must take account of all the substitute forms of signaling, because a change in the legal penalty for violation of the marriage vow will not make it easier or harder for parties to capture the marital surplus if near substitutes are available.

The Relationship Stage

Nonlegal Enforcement by Husband and Wife

The marital relationship can be modeled as a repeated prisoner's dilemma, in which each spouse deters the other from cheating by threatening to punish any transgression of the marital obligations. One strategy might be tit-for-tat: each spouse "cheats" in response to cheating by the other spouse in an earlier round. The husband fails to come home on time; the wife refuses to agree to the vacation he wants. The wife withholds some of her income from the household; the husband starts staying out late with friends. To avoid retaliation, spouses do not cheat. Information problems complicate the interaction, however. Parties may disagree about what counts as cheating, or whether an instance of cheating really occurred. Discussions, arguments, and threats may resolve these issues prior to any real punishments. From these interactions the marital obligations emerge (recall that they cannot be specified in advance), always subject to change in light of new circumstances.

Spouses mitigate information problems by signaling. Ritualized gift-giving and celebrations are typical. Another way for spouses to mitigate information problems is by dividing obligations and rewards in certain ways. Suppose a husband and wife jointly work and jointly take care of their child in shifts. The problem is that each partner cannot observe and evaluate the actions of the other. One spouse does not know whether the other works hard enough in the market and cares for the child with adequate attention. If the husband is fired, he can blame the firing on his boss' unreasonableness when in fact it was his own shirking on the job. He did not work hard because he knew that he could rely on his wife's wages in case he were caught shirking and fired. If the husband watches television with his child rather than taking care of the child in a more active way, the wife may not observe the effects of this inattentiveness for years. This moral hazard problem affects both parties: each may shirk both in household and in market production because he or she shares the residual with the other spouse.

The traditional family may be considered a solution to this joint moral hazard problem, as Scott and Scott (1998) suggest.[4] The wife specializes in household production, and the husband specializes in market production. The wife does not shirk because she bears the full consequences of bad parenting: if the child is in a bad mood the next day because she put the child to bed late, then she, not her husband, bears the consequences. The husband does not shirk at work because if he is fired, he cannot rely on his wife's wages. In addition, the husband and wife specialize on the benefit side, not just on the cost side. The child has an affectionate relationship with the mother and a distant relationship with the father. The wife thus obtains the full residual from good parenting. The father obtains all the status and money from his market production beyond the costs of supporting the family; thus he obtains the full residual of his market work. Cooperation, then, is limited to sharing, not production. These considerations might account for the relative stability of the traditional family over time and its appearance in different cultures (though not the usual assignment of household obligations to women), and also the current problems with the ideal of perfect equality between the sexes, where both participate equally in the market and in household production.

But the traditional family is highly unsatisfactory in the modern era, and one can detect new responses to the moral hazard problems that do not condemn spouses to traditional roles but that also do not conform with the ideal of perfect equality. The armchair anthropologist of the mating rituals of American middle-class young adults will detect three trends: the rise of the "sensitive male," the increasing respectability of part-time or low-intensity careers for educated women, and the gradual decline of the stigma against the "house husband." The last two trends suggest a return to the division of labor in the household, but less extreme than before and less dependent on traditional sex roles; thus, spouses recapture the advantages of the traditional marriage, where each has more or less exclusive authority over one aspect of the household, and both incur the costs and enjoy the residual benefits that are associated with his or her sphere of authority. The first trend suggests that men are promising in return to take more childcare responsibility and have a closer relationship to their children. Women insist on such promises from men more than they did in the past; because women incur higher opportunity costs from staying home than in the past, they want men to commit to some childcare, and altruism commits men to engage in some childcare. Naturally altruistic men, who had to suppress their altruism under the old system, profess their sensitivity and no doubt demonstrate it during the courtship; and now less altruistic men must mimic the more altruistic men during the courtship stage in order to avoid rejection. Having committed themselves to be sensitive males ex ante, even non-altruists may suffer reputational sanctions if they fail to be sensitive ex post. It is not clear whether this new arrangement will be stable: there are advantages and disadvantages to both sides.[5]

So information problems and strategic opportunities within a relationship, and the high-value opportunities outside a relationship, all pose a threat to the stability of marriages. To protect themselves, men and women engage in signaling at the courtship stage, and husband and wife take on roles that minimize information problems while allowing each to deter opportunism through the threat of retaliation. However, there are additional nonlegal sanctions, beyond those in the power of husband and wife, that help deter opportunistic behavior.

Nonlegal Enforcement By the Husband's and Wife's Families

Historically, the parents of husband and wife have played a larger role in deterring opportunistic behavior than they do in modern American society. This was true both at the search stage, in which the families might arrange marriages or at least cull "respectable" mates from the pool; and at the relationship stage, in which the parents of one spouse would provide protection for him or her, and would retaliate against the other spouse when that person engaged in opportunistic behavior. Most significantly, parents can influence the behavior of their children and their spouses by distributing wealth to the household or withholding it. It remains true today that parents sometimes threaten to refuse to pay their children's college tuition if the children do not break off relationships they do not like (often on religious grounds). (Striking examples of such parental influence can be found today in India.)

At first sight, one might wonder what is added through the parents' participation. It might appear that instead of having a repeated game between the relatively poor husband and wife, one has a repeated game between the relatively wealthier parents of each. But increasing the endowments of two players in a repeated prisoner's dilemma does not, by itself, increase the chances of cooperation, and may reduce them.

One advantage of parental involvement is that parents have better information than their children about potential mates, including their wealth, their skills, the advantages of their family connections, and their reliability. Parents usually have better judgment, as a result of age, and it is not clouded by emotion. They might foresee future conflicts overlooked by their children, such as the importance of religious differences when it is time to educate grandchildren.

Another advantage of parental involvement arises when the spouses have unequal endowments: the parents of the poorer spouse can provide leverage in household disputes. If one spouse has greater outside opportunities, the other spouse might not be able to deter cheating through tit-for-tat and similar strategies; but if the latter spouse's parents can bring pressure on the first spouse, the relationship can be maintained. A person might be deterred from cheating by the prospect of gifts and an eventual bequest from the spouse's parents.

A more interesting advantage of parental involvement is that parents have longer time horizons than their children do when the children are at a marriageable age. Most adults invest in a reputation as responsible members of the community. The family name becomes a kind of trademark associated with reliability. Parents might want to transfer this reputational advantage to their children for several reasons: because they care about their children, because they want their children to be wealthy enough to support them when they are old, and because they want their children to maintain the family name. Indeed, the "family," like a corporation, might be thought of as an immortal (or at least indefinitely long-lived) institution that solves the last period problem by giving every member an incentive to maintain the family reputation (compare Kreps 1990a). Parents try to restrain their children's romantic and sexual impulses, and encourage them to make marriage choices based on the wealth and status of the potential partner.

The disadvantages of parental involvement are numerous. History, theory, and common experience contradict the common assumption that parents' and children's interests are aligned. Parents care about themselves *and* about their children. Otherwise, few parents would retain leverage over their children by retaining wealth until their deaths, and would instead purchase an annuity and give their children, while still young adults, the excess over what the parents will need for the remainder of their lives. Parents might be more concerned about the way the family name supports their interests than their children's relationships. By promising bribes or threatening punishment, they might force their child to marry someone who is not a good partner but who would be a good ally for the parents. The tension between parents and children in this respect culminated in the frequent practice in early modern Europe of clandestine marriages, and, since then, elopement.

Parents frequently try to educate or prepare their children so that they do well at the search stage, without taking account of how well they do during the relationship. This may be due to the fact that parents obtain payoffs from their children's relationship at an earlier stage in that relationship than the children do. A "good marriage" redounds immediately to the benefit of the family name, leaving the children to bear the consequences of a possibly loveless relationship. For example, parents have frequently kept information about sex from their daughters in order to suppress their curiosity and deter experimentation prior to marriage. This allows the daughters to enjoy reputations as good types during the courtship stage, but it leaves them ill-prepared for the sexual demands of the relationship.[6] The practice of clitoridectomy is sometimes similarly justified on the ground that it makes women more desirable marriage partners; it too may sacrifice the emotional (as well as physical) well-being of the daughter to the interests of her parents.

Nonlegal Enforcement By the Community

Nonlegal enforcement by the community is, historically, of considerable importance. The most striking examples come from a cluster of practices often called "charivari" which occurred in virtually every Western European culture (and no doubt others as well) in every historical period up to the 20th century. Charivari, greatly simplified, consisted of gangs of local youths and young adults who would harass people whose behavior violated local family norms, including norms against marriages between people of different ages or status, against wife-beating and husband-beating, and against adultery.[7] The harassment would typically last a day or a few days; it would involve festive or carnival elements, including parades, noisemaking, the wearing of costumes, and games. Sometimes the gang would make noise in front of the offenders' home; sometimes it would parade around an effigy, perhaps burning, mutilating, or burying it in a mock funeral; and sometimes it would force the offender to engage in some humiliating public action, like riding an ass backwards around town, or enduring members of the gang pelting him with mud and excrement. Authorities generally tolerated this behavior before powerful governments came into existence, partly because they saw its advantages and partly because they were too weak to suppress it.[8]

The model in Chapter 2 can be used to analyze community enforcement. Just as potential mates signal to each other their reliability as cooperative partners, so do members of the community signal to other members their reliability as cooperative partners. One signal of reliability is participation in (costly) nonlegal punishment of deviants who threaten or appear to threaten the community. The problem of free-riding is partly overcome by everyone's desire to reveal that he or she is a cooperative type. Notice that people will also marry in order to signal to the community (not just to each other) that they are reliable types.

If formal marriage serves to deflect suspicions about one's reliability as a member of a community, then in equilibrium people might believe that those who marry by a certain age are good types and those who do not marry by a certain age are bad types. Their attitudes will be revealed in many ways, some subtle, some not. The unmarried might be less sought after as business or social partners; they might be considered "defective" in a vague sort of way. To avoid these inferences about themselves, everyone or almost everyone might marry, even if expecting the marital surplus (as defined earlier) to be negligible.

The desire to signal may result in the production of external public goods. If people who have illegitimate children cannot support these children, and they become burdens on the community, people will resent those who have illegitimate children. They may signal their type by ostentatiously avoiding such parents despite the gains that could be obtained by cooperating with them. They might forbid their own children to play with the ostracized children. These sanctions may deter people from having illegitimate children in the first place, or may cause them to take responsibility for them by marrying their partners and setting up a normal household.

Such community enforcement—which arises endogenously from people's desire to enhance their reputations—becomes less effective when people can easily exit the community, or it is difficult to observe people's behavior, or to communicate detected transgressions. It becomes less effective when members of the community have heterogeneous values. All of this is familiar. The more interesting point is that the signaling analysis suggests that individuals interested in showing their loyalty to the community will sanction anyone who can plausibly, or even just momentarily, be thought a threat to the community. One way to draw attention to one's attack on an alleged deviant is to attack in an entertaining way, for if other people do not observe one's punishment of a deviant, this action does not serve as a signal of one's loyalty to the community. This is why charivari involved mockery, parody, music, gaming, and other festive elements. These elements drew observers and participants, overcoming the problem of collective action. But if punishment is fun, or if the person who punishes receives rewards, it may be initiated for its own sake, not because the victim deserves to be punished. Gossip and ostracism are directed against anyone who behaves in an unusual way, and through feedback effects people who are not guilty of any harm may be severely punished. Signaling can lead to spiraling violence, such as lynching (Ingram 1984), and it can be exploited by bad types out for private gain (Ingram 1984, Wyatt-Brown 1982), including those who feel resentment toward good marriages.

Communities have less control today over the family relationships of their members than in the past, but they still have much control. Politicians pay a heavy price when they violate ordinary family norms, and no doubt less visible people do as well. Ordinary workplaces all have their tales of sexual intrigue and exploitation. (Law school charivaris—the skits put on by students to mock boring and inept law professors—sometimes spill over to attacks on their personal and family lives.) Informal systems of social control always exhibit such pathologies.

Legal Implications

The discussion so far shows that there are many sources of nonlegal constraint on marital opportunism, and each source produces pathologies as well as advantages. Notice that social norms are endogenous in games in which people maximize their interests through cooperation. They do not exist independently of people's interests, beliefs, and behavior. The question now is whether the analysis of social norms helps one understand the law of the family.

Divorce. Scott and Scott (1998) argue that a hypothetical marital bargain—which describes the marital obligations to which parties would agree if they had perfect information ex ante—would restrict divorce. For a traditional family, where H works and W raises the children and takes care of the home, a termination rule might require a two-year waiting period after which H must pay alimony to W. The two-year waiting period deters W from opportunistically suing for divorce, as does the limitation of alimony to, say, support. The alimony requirement deters H from opportunistically suing for divorce. By making the divorce costly, the law gives each party a powerful weapon for retaliating when the other acts opportunistically. For a dual-earner family, an alimony rule makes little sense, but the two-year waiting period could deter some opportunistic divorces. Notice that the law does not itself punish the opportunist: by hypothesis, courts cannot verify fault.

This argument recalls the signaling model. H commits himself to W by marrying her, because if he subsequently divorces her, he incurs significant costs. If H were not interested in W or had a high discount rate, he would not send this signal. But if H and W anticipate a dual-earner family, the two-year termination provision would hardly seem to impose much of a cost on either party. H and W can both pursue intimate relationships with other people during the termination period; they can even have, with these other people, children no longer stung by the stigma of illegitimacy as in the past; they are only restricted from a subsequent formal marriage within those two years. Yet it is quite possible that H and W face as deep a commitment problem ex ante as they would in a traditional household. So in the dual-earner scenario the signal is too cheap, and ineffective.

One response to this argument is that the law should make available a menu of options to the parties. If a two-year waiting period is insufficient, then the parties can choose a five- or ten-year waiting period. The logic of the argument has no natural stopping point, however. Why shouldn't the parties be able to choose a no-termination rule? Some might say that parties are not competent enough to choose such a rule. The competency argument, however, explains too much: why should we believe parties are competent enough to choose a five- or ten-year rule? Lacking any principle for distinguishing terms only incompetent people would choose from those that competent people would choose, one could justify any restriction on the marital contract or none. But competency to choose terms is not the issue, anyway. Competent parties could agree to a contract that said that upon divorce each party had to pay half his or her future lifetime wages to the government or to a charity. They could no doubt agree to stranger contracts: marriages in which, say, only spousal abuse but not extra-marital affairs are grounds for termination; or marriages in which one spouse is permitted two illegitimate children and the other spouse is permitted none. They could agree to a marriage contract anticipating a traditional household except that the woman must pay the man if he decides to divorce her. Any of these contracts could be chosen by competent parties who seek to overcome commitment problems.

Rather than supplying default rules and allowing parties to contract around them, marriage law offers a basket of immutable obligations and forbids almost any deviation from it. The long history of restrictions on marital contracts includes restrictions on sex before marriage and outside of existing marriages (including laws against fornication, adultery, and bigamy, and penalties inflicted on illegitimate children); mandatory rules governing divorce, distribution of property, and authority over children; and rules governing the authority within marriage over property. What is puzzling is not the content of the rules. It is the fact that they are not default rules, but, for the most part, mandatory. The puzzle that must be solved by a theory of marriage law is why people are not allowed to contract around the laws of marriage.

I do not have a fully satisfying explanation for these phenomena, but let me suggest a few possibilities.

1. The restrictions on freedom of marital contract make the formal marriage contract *focal,* facilitating community enforcement. Imagine that there were no restrictions, and I could have four "marriages" with four women, who perhaps are also married to other people, and I could have children with all four women; or in a single marriage my spouse and I could agree that she may have two extramarital affairs and I may neglect the children. Put aside the interests of the children for the moment, and suppose that one can credibly commit to any one of the relationships in the first place only if community enforcement deters opportunism. The problem is that the existence of multiple or idiosyncratic relationships might be so confusing to the members of the community that community enforcement becomes impossible. Whereas an extramarital affair counts as opportunism in a traditional household, and

quickly and easily calls for community sanctions, it does not count as opportunism in my household if my spouse and I have permitted a certain number of affairs in our contract. The members of the community cannot be expected to know enough about our marriage agreement to be able to determine whether any particular action violates the agreement. They need a standard form marriage against which to measure mine, so that they know what counts as opportunism. Marriage can serve as a signal of commitment only if community members deter opportunism, but community members can deter opportunism only if the actions that count as opportunistic are common to all marriages.[9]

2. The restrictions on freedom of marital contract, by making the formal marriage contract focal, facilitates *legal* intervention. The purpose of requiring a marriage license and standardizing the marriage contract is to establish certain verifiable actions as a violation of the marriage contract. Under the old fault system, adultery but not emotional coldness constituted grounds for divorce. In a real marriage it is quite possible that an instance of adultery would be forgivable but emotional coldness over the long term would not. Still, one might argue that *on average* adultery constitutes a violation of the marital bargain whereas emotional coldness does not, or that adultery cannot usually be deterred by the victim through retaliation whereas emotional coldness usually can. If so, legal punishment of adultery but not of emotional coldness would deter opportunism more effectively than legal deference to all forms of marital opportunism.[10]

3. The restrictions on freedom of marital contract prevent parents from abandoning or neglecting their children. The problem with extra- and premarital affairs, and with divorce, is that they may result in children whose parents (one or both) prefer starting a new household to taking care of them. Restrictions on the marriage contract do not, on this theory, necessarily serve the interests of the parents; they serve the interests of children or authorities who may be responsible for supporting unwanted children.[11]

4. The restrictions on freedom of marital contract promote equality in the marriage market—either between men and women, or between high-status and low-status people of the same sex. Consider restrictions on formal polygamous marriages and on de facto polygamy (for example, men keeping mistresses). In polygynous relationships women must share the marital surplus with each other and with the man, resulting in great inequality. Polygynous relationships also deprive low-status men of wives. Some historical evidence suggests that restrictions on such relationships arose in part at the behest of civil and religious authorities who were trying to win the support of ordinary people. The latter resented the powerful aristocrats who took concubines from the pool of marriageable lower-class women. The contrary assumption

that the purpose of family law is to enable parties to maximize the marital surplus prevents one from accounting for laws that are designed to equalize shares of the marital surplus rather than to maximize them.[12]

5. Strict divorce laws might have been justified at a time when parents maintained great control over whom their children married. Suppose that children at marriageable age have poor judgment and lack the self-control needed to overcome sexual passion and to evaluate a potential partner soberly. If divorce and cohabitation laws are strict, people will make and be unable to leave bad marriages. But if parents control their children's marriages—perhaps through a legal consent requirement or through their control over the children's future wealth—then they may ensure that children make desirable marriages. Strict restrictions on divorce and extra-marital relationships, then, might serve as a commitment mechanism after the parents' influence wanes. An analogy can be found in the restrictions on loans to expectant heirs, who are tempted to use their expectancies as collateral for loans that finance their youthful dissipation. However, marital restrictions might also have reflected non-altruistic efforts by parents to maintain control over their children.[13]

Notice that each theory justifies a strict divorce law but also requires restrictions on near substitutes for legal marriage. If the termination rules are too severe, people will not marry but cohabit and have illegitimate children,[14] or they will have adulterous relationships. Each theory therefore justifies laws against pre-marital fornication, adultery, and bigamy, and also laws punishing illegitimate children as a means of deterring extra-marital sex. In addition, the first two theories explain the refusal of the law to enforce idiosyncratic sharing arrangements and obligations within marriages. The third theory may explain restrictions on inheritance and gift-giving.

The recently enacted Louisiana covenant statute is unlikely to have a substantial effect on marital arrangements, nor would a more aggressive statute that provided a menu of options, as advocated by Scott and Scott (1998). Unless the legislature also decides to crack down on adultery and illegitimacy, "covenant marriages" will not be significantly more costly than ordinary marriages, because parties can substitute to informal marriages (that is, adulterous relationships) if their covenant marriage becomes unsatisfactory. One might argue that the community will punish people who violate covenant marriages, but community enforcement depends on clear signals. Even if the community were willing to punish people who violate covenant marriages, it could do so only if it knew whether a particular marriage was a covenant marriage or not. But people with casual knowledge of the marriage will not know what kind of marriage it is, and there is no opportunity for a court-like proceeding in which the victimized spouse proves to the community that the marriage was a covenant marriage. For community enforcement, there would have to be different kinds of wedding ceremonies establishing different levels of commitment, with each ceremony being different enough to make an impression on observers and to allow them to remember the level of commitment. But probably simply knowing whether a couple is married or not exhausts the recollective capacities of the public. Because covenant marriages will be no more binding than ordinary marriages, they will not be more effective signals of commitment than ordinary marriages.

Intra-marital disputes. Scott and Scott (1998) argue that the hypothetical marital bargain would forbid the legal enforcement of its terms because (a) courts cannot determine whether actions within the marital relationship constitute opportunism or not, and (b) social and relational norms deter opportunism.

This argument, however, is not valid or it is not complete. Suppose the law gives the wronged party the right to sue the other spouse. Initially, if husband and wife enjoy the marital surplus because their use of tit-for-tat or a similar strategy deters marital opportunism, then it does not matter whether the law gives one spouse an obligation against another or not. Because nonlegal resolution of any dispute would be cheaper than legal resolution of any dispute, a lawsuit would count as opportunism, which would be deterred by the threat of retaliation. The legal enforceability of terms is *irrelevant* to spouses who successfully cooperate. But this does not mean that potential mates would prohibit legal enforcement of marital terms. Suppose that W cannot deter the opportunism of H through the threat of retaliation, because she cannot withhold an amount of the marital surplus that exceeds the value to H of his opportunistic act, and at the same time withdrawal from the relationship would be exceptionally costly for her. The ability to sue H might give W a credible threat of retaliation that would deter H's opportunism in the first place.

This could happen in many different ways. First, if courts *could* verify bad behavior, even if not terribly accurately, then W's suit might succeed; H, in order to avoid legal costs, would cease his opportunistic behavior as soon as W threatened to sue. Suppose, for example, that W threatened to reveal H's adultery to the authorities in a legal system in which adultery is a crime. If adultery would usually but not always constitute breach of the marital contract, a court that routinely punishes an adulterer would, more often than not, correctly enforce a term of the marital contract.

Second, even if courts could not verify bad behavior and even if an actual lawsuit would injure W as much as it injures H, W might have a credible threat to sue if by doing so she invokes community sanctions against H. Members of the community might not believe W's tales of H's adultery until she brought a lawsuit, for the lawsuit, because it is costly to W, is a credible signal of her injury.[15] This would be true whether or not the members of the community could independently verify H's opportunism.

One cannot always depend on the community to assist in the maximization of marital surpluses. Communities (and parents) have frequently disapproved of various unions, especially unions between members of different races, religions, and classes, and exerted pressure on the spouses to end their relationships, even though these marriages produce marital surpluses for the spouses like any other. Communities have frequently disapproved of unions where the wife dominates the husband or the spouses do not demand sexual exclusivity from each other or, even, in some cultures where the husband does *not* have a mistress. Men might engage in adultery because members of the relevant community admire those who display sexual vigor or who have enough wealth to support multiple households; but the adultery nevertheless violates the terms of the value-maximizing marital contract. When community pressures conflict with the norms of the relationship, there is no reason to believe that "social and relational norms" in toto deter marital opportunism more effectively than the law does. Perhaps the law gives the parties the power to resist destructive social norms by enabling one spouse to retaliate against the other if he succumbs to them.

The treatment of children: the waning stigma of illegitimacy. Until recently illegitimacy was highly stigmatizing in the United States as well as in other countries. The stigma against illegitimate children has been reflected in laws that exclude illegitimate children from offices, restrict their right to inherit, and impose other disabilities. Today, both the legal and the nonlegal disabilities have almost evaporated.[16] There was, in the United States, a transitional period in which the Supreme Court struck down laws that discriminated against people on the basis of their illegitimacy.

Why have people discriminated against illegitimate children? The practice is puzzling, because it violates powerful moral commitments against treating blameless people as instruments for punishing others. The story might be something like this: Community members are always looking for ways to signal their loyalty to each other. Actions which are unusual or that pose a threat to the community present an opportunity to demonstrate loyalty. Households with illegitimate children present such a signaling opportunity, either because they violate a statistical norm (most people have children only with a single spouse) or because they result in unsupported children that become a burden for the community. People signal their condemnation of this behavior by avoiding both the guilty parties and the products of their actions, the illegitimate children. This behavior closely resembles the practice of discriminating against innocent members of a minority which is widely blamed for problems suffered by the majority (see Chapter 8). In addition, parents of adult children who have illegitimate children might neglect the illegitimate children because they otherwise interfere with efforts to preserve family wealth intact. And spouses might sanction each other for engaging in adulterous relationships by neglecting the children of those relationships. The illegitimacy stigma, then, arises because people pursue a strategy of sanctioning the children of illegitimate relationships as a way of punishing the children's parents in a repeated prisoner's dilemma or of signaling their loyalty to the community.

This might explain why the illegitimate children of wealthy and powerful parents often do not become stigmatized. The community does not stigmatize the child because its members do not want to lose the patronage of the parent or, for that matter, the child when he grows up.

The illegitimacy stigma poses a challenge to those who see social norms as a benign influence on marital relations. To be sure, the stigma *might have* discouraged the production of children who could not be cared for and the initiation of sexual relationships that caused conflict and disorder. But the model suggests that the stigma arose as a side effect of other concerns. It created a pool of stigmatized people who could not cooperate on equal terms with other members of society. The rapid decline of the stigma over the last few decades suggests that even if the illegitimacy stigma once served a purpose, it has long outlived that purpose. The historic rise of a *public* family law, then, might be seen as an effort by authorities to substitute neutral legal enforcement mechanisms for more chaotic discriminatory nonlegal enforcement mechanisms.

Same-sex marriages. Advocates of legal recognition of same-sex marriages can appeal to libertarian and free market traditions in this country, but they run up against a stubborn insistence that same-sex marriages would subvert important values. No one has explained what these values are and why they are worth defending. Opponents leave the implicit religious arguments unarticulated because they do not count for much in educated debate. Some opponents say that same-sex marriages would subvert the institution of marriage, and that marriages produce important social benefits, but they do not describe the process by which this subversion would occur.

A possible approach to this complex topic is suggested by the argument that the legal recognition of same-sex marriages might change the "meaning" of marriage in an important way.[17] The meaning of marriage might be understood to refer to the inferences that observers draw about people who are married. To see what these are, consider traditional views about unmarried people. Traditionally, unmarried men and women have been stigmatized, though in different ways. The stereotype of the elderly bachelor was that of someone who does not take his responsibilities seriously, who likes to have a good time, who is somewhat undisciplined. The stereotype of the elderly spinster was that of someone who lacks passion, who is cold and perhaps a little selfish, who is a bit dotty. Thus, by entering marriages people avoided these unpleasant inferences about their characters. These stereotypes are not as strong as they used to be, but they help reveal the meaning of marriage, as it used to be and as it is today. People who marry are regarded as generally trustworthy people—self-disciplined people who can make commitments and keep their word—albeit people who have rather conventional tastes.

If this is the meaning of marriage, then it is hard to understand why people would object to same-sex marriages. Indeed, advocates of same-sex marriage emphasize the disciplining effect of marriage (Eskridge 1996), and this claim may be interpreted, in part, as an argument that if the signal of the legal marriage vow were available to homosexuals, they would, like heterosexuals, have an easier time entering long-term monogamous relationships.

One objection might be that extending recognition of marriage to a greater number of people benefits those people at the expense of those who were entitled to the recognition under the narrower definition. To understand this objection, suppose that romantically involved people who do not live together, or who have multiple boyfriends or girlfriends, declared that they are "married" to whomever they are involved with. They might go through formal ceremonies, and obtain rings and legal rights. One possible result is that people would become confused about how to treat the people who declare themselves married but who do not act like ordinary married people and do not obtain marriage licenses. They might become confused about who is really married and who is not really married. To avoid embarrassment, people might treat these fictive marriages as real marriages. As a result, the pool of "married people" would increase from those pairs who have licenses and live together, and so on, to all those who are romantically involved in a certain way. Then the value of being identified as a "married person" will decline for those who are traditionally married, because the same label is applied to people who are not able to commit themselves to long-term monogamous relationships as well as to people who are able to commit themselves. If commitment to such a relationship is a signal of a person's type, the marriage vow becomes a weaker signal of type than it had been before. Having invested in the marriage vow, the "really married" would suffer a decline in the value of their reputations—unless of course they could arrive at a new signal that would distinguish them from the merely romantically involved. Living together over a long period of time would serve as such a signal, but could not be used by people entering a relationship. Anticipating the dilution of the meaning of their marriages, married people would oppose any effort by individuals to expand the definition of marriage. This is like expanding the group of soldiers who are entitled to a medal for heroic behavior. Those who already are entitled to the medal under the narrow definition oppose the redefinition of heroism because it dilutes the status that the medal confers on them.

This argument applies to same-sex marriages. Permitting same-sex marriages would, by influencing the way we make judgments about those who are known to be married, reduce the value of that label for those in traditional marriages. Thus, heterosexual people fear that if same-sex marriages were recognized, their marriages would be less effective signals than they already are. But there is an additional premise here that same-sex marriages would involve less commitment than ordinary marriages. This is possible, if only because concerns about the well-being of children, which are likely to be more common in heterosexual marriages than in same-sex marriages, bond partners together. But it is not clear whether this premise is true, and if it is true, how important it is in the argument.

The "privatization of family law" is said to consist of the increasing replacement of status with contract (ignoring the fact that medieval marriages were far more contractual than modern marriages). Let me suggest a more literalistic interpretation, namely, the increasing exclusion of the crowd from participation in the regulation of the family. There are two issues here: (a) whether people should be able to enter and have enforced idiosyncratic marriages; and (b) who should enforce those marriages. We might condemn the historic involvement of the community and even of parents in their children's marital choices, on the ground that sometimes dysfunctional mechanisms for promoting the solidarity of the community result in hardship for people with idiosyncratic tastes and needs. Fear of the crowd has always induced conformity, and although the prospect of shame and ostracism have discouraged some obvious breaches of marital etiquette, such as desertion and abuse, there is no reason to believe that it maximizes the value that people obtain from marriages. The gradual replacement of social and parental sanctions with legal sanctions may reflect a reasonable preference for the order and precision of the judicial bureaucracy over the pathologies of the crowd.

If enhancing the effectiveness of community enforcement of marriages is, even if possible, undesirable, then one might think that the only alternative is to increase the strictness of divorce laws. But making divorce law stricter, whether or not at the option of the parties, will not by itself increase marital surpluses. Stricter divorce laws will not necessarily cause people to engage in more careful searches and more binding commitments; instead, such laws would likely cause people (ex ante) to delay marriage, cohabit, and produce illegitimate children, or (ex post), having married, to enter adulterous relationships and produce illegitimate children. To change behavior substantially, the state would have to launch a full-scale assault against all the near substitutes of marriage, and this would mean bringing back archaic penalties on fornication, adultery, illegitimacy, and serial polygamy. Given the indifference to these behaviors among a large sector of the population, and the powerful interests that support them, a successful legal intervention—one that encouraged marriage, deterred cohabitation and illegitimacy, discouraged divorce, and maximized the marital surpluses—would have to be dramatic and highly coercive. This would be neither desirable nor, as long as current trends continue, possible.

6

Status, Stigma, and the Criminal Law

A recent wave of interest in shaming penalties raises questions about the ways in which criminal punishments deter people from committing crimes. The traditional economic theory of crime assumes that punishment deters because it reduces utility (Becker 1968), but it says nothing about *how* punishment reduces utility beyond the obvious points that fines reduce wealth and that prison terms reduce the opportunities to earn wealth and to spend it on preferred goods. Advocates of shaming penalties argue that public humiliation of the offender will, in some circumstances, deter criminal behavior in a highly cost-effective manner. Those who commit minor crimes, like paying for sex or drinking in public, should be required to wear signs proclaiming their offense, or to place bumper stickers on their cars, or to stand in a public place for a time, or to wear distinctive clothing; and those who commit serious crimes should be required to spend time in prison, rather than paying a fine, because imprisonment stigmatizes and paying a fine does not. The point of publicizing the crime in these ways is not so much to persuade the public that criminal behavior will be punished, or to inform it about the penalties for certain crimes, or to inform it about which behavior is criminal. All of these goals are consistent with the traditional model or can easily be absorbed by it. The point of publicizing the crime, according to the shaming advocates, is to cause the public to avoid or ostracize the offender. The point is to *shame* (Kahan 1996, 1997).

Shaming penalties have a long history. Looking around the world and back into time, one can compile a very long list of them, which would include the following: flogging, branding, tattooing, distinctive clothing requirements like the famous scarlet letter, public apology, amputation of limbs, hideous execution, confinement in the stocks, blinding, and ritual humiliation, such as a mock beheading (Spierenburg 1995, pp. 53–54). Notice that many of these punishments are clearly designed to communicate the offender's status to the public. Others, such as flogging, can be performed outside of public view and often were, but they were also often performed in public view. Public flogging was a different punishment from private flogging, just as public execution was different from private execution. Contemporaries often referred to the shaming effect of these punishments, and this suggests that the punitive impact resulted in part from the fact that people would treat the offender, or his family, differently after seeing him shamed.

Advocates of shaming penalties have not argued for bringing back the pillory, but they do argue that lesser forms of shaming are desirable punishments, because they are cheap and they have a powerful deterrent effect. The average member of the middle class would be less likely to buy sex from a prostitute if the punishment were public exposure rather than a fine. The average white collar criminal fears the shame of a prison term more than a fine, which is not so shameful. These arguments rest on the assumption that people care deeply about their reputations; therefore, the government can exploit this concern and the fact that reputations do matter in daily life by destroying people's reputations rather than punishing them in a traditional way.

These arguments provide an opportunity to examine the implications of the cooperation model for criminal law.

A Model of Shaming

Although shaming penalties can be imposed by the government, shaming is essentially a nonlegal sanction. Shaming occurs when people draw attention to the undesirable traits or behaviors of another person, with the result that the target is seen as a less desirable cooperative partner. People might simply avoid the target thereafter, or not trust him, or trust him less than before; or they might taunt or scold the target. But taunting and scolding are effective only if backed up by the prospect of real ostracism. The sting comes from the expectation that because one has developed a reputation as a bad and unreliable person, one will lose future opportunities which have significant value.

The act of shaming a person, then, should be distinguished from a relatively passive refusal to deal with a person who has developed a bad reputation. Shaming a person is an entrepreneurial activity: it requires time and effort, and the reward is uncertain. It takes time and effort to talk to members of one's community and to argue in a persuasive way that the victim has engaged in bad behavior. And this behavior involves risk. The target of the shaming, and his friends and family, might retaliate against the speaker by taking violent action against him or by spreading unflattering rumors about him. If the shame entrepreneur does not persuade his audience, they may conclude that he is a busybody or a deceiver, a bad type who ought to be shunned. For these reasons, shaming a person is usually understood as a signal by the speaker that he belongs to the good type.

The first person to engage in public shaming may have inside information or he may have the most to gain by signaling that he belongs to the good type. Once this shame entrepreneur sends the signal, listeners must respond. If they believe him, they may respond by merely shunning the target of the shaming. But they may also seize the opportunity to signal as well. An individual might fear that if he does not engage in an act of public shaming, people may believe that the signal is too costly for him. The signal could be costly either because this individual values his interactions with the bad type, or because he discounts the future highly. Both possibilities will diminish his reputation. As more people signal their type by shaming the victim, the incentives to join in will increase, because of the fear that observers will assume that the dwindling few who do not join are the worst people of all.

It is worth stressing that the model provides two reasons why a third person will avoid dealing with someone who has been shamed. (1) The deviant has been revealed as a bad cooperative partner; the third person would rather enter a relationship with someone else whose reputation is unspoiled. (2) The third person fears that if he fails to sanction the person who has been called a cheater, others will sanction him (the third person) by avoiding him. That is, the third person seeks to signal that he belongs to the good type by participating in the sanctioning. Only if people are motivated by the second reason will there be a cascade, resulting in the extreme public humiliation of the deviant and even his lynching at the hands of a mob. The mob is a collective disaster produced by individually rational behavior, much like a bank run.

Shaming Penalties within Subcommunities

Consider a subcommunity that is regulated by a legal system controlled by the members of a larger, dominant community. Some members of the subcommunity engage in deviant behavior that is not deterred by the law. The law might not deter this behavior because the kind of deviant behavior is difficult to prove in a court, or because police officers and other legal officials focus their energies elsewhere. Because the law provides no deterrent, any deterrent must come informally from the subcommunity. At the same time, it may be difficult for members of the subcommunity to erect and enforce a law-like system, which would involve compulsion. Although the police might not bother to arrest prostitutes in the subcommunity, they might arrest members of the subcommunity who try to punish the prostitutes, or who form vigilante groups that enforce the law in a systematic way.

Under these circumstances a cohesive subcommunity may or may not be able to maintain order. Some individuals may expose and punish deviants in order to signal their type, and if others respond either by actively shaming these deviants or passively avoiding them, it is possible that predatory behavior will be deterred. Informal sanctions might include scolding the deviant, which may be intrinsically painful or humiliating but also is a warning that more serious punishment will come if behavior does not change; harassing the deviant, perhaps physically, but in a minor way that would not draw the attention of the criminal justice authorities; ostracizing the deviant, which would include refusing to employ or work for or socialize with or do business with the deviant; and even murdering the deviant or harming him or his family. Hawthorne (1981, p. 51) says of the effect of the scarlet letter on his heroine—"It had the effect of a spell, taking her out of the ordinary relations with humanity, and enclosing her in a sphere by herself." None of this is foreordained, of course. Two plausible alternatives to this decentralized system of enforcement are, first, that one or a few people accumulate power rapidly, and eventually, in order to consolidate and protect their gains, keep the peace; and second, that people rely heavily on their families for protection. The first alternative is familiar from the phenomena of organized crime and gangs,[1] the second is familiar from the history of feuds among families under conditions of limited government.

The advantages of this system of informal sanctioning are that it exploits people's ability to monitor each other, their networks of communication, and their desire to avoid people who are low types.[2] The marginal cost of monitoring and communicating is trivial, given that people already must constantly interact with each other for social and business reasons. Accordingly, informal sanctioning does not require large capital investments in police, judges, and prisons (which may be out of the question when the criminal justice system is operated by an unsympathetic government). These advantages are powerful enough that informal systems of social control always coexist with formal criminal justice systems, because the criminal justice system does not deter criminal behavior that is difficult for authorities to detect or punish, or that occurs within subcommunities that have little political influence.[3]

But the disadvantages of informal sanctioning are also significant. A familiar problem is that the informal system does not work when subcommunities are large and the value of people's outside opportunities are high. Another problem is that the system does not exploit economies of specialization. These problems create significant pathologies.

Pathologies of Shaming

Optimal deterrence occurs when the community imposes a sanction on the offender that injures him by an amount equal to the expected cost that the act imposes on others. If a store loses $100 from shoplifting, then the punishment should equal $100 divided by the probability of detection.

Suppose an individual learns that someone is a thief. The person will expose the thief if the gains exceed the costs. The costs include the possibility of retaliation from the thief, plus the time and expense of revealing the thief's behavior. The gains include the possibility that others will regard the individual as a good cooperative partner since he has incurred the cost of exposing the thief. It is clear that there is no correlation between the private gain from exposing the thief and the social gain. The individual's decision whether to expose the thief will depend not on the cost imposed by the thief on the victim, but on such things as whether the individual's reputation will be enhanced; but this will depend on a variety of extraneous factors, including whether the individual's job gives him returns from having a cooperative reputation, which is more the case if he is a politician or businessman than if he is a craftsman or novelist, and on people's prior beliefs about the distribution of types.

Further, the cost to the thief of being exposed is uncorrelated to the social cost of his behavior. The thief's reputation might matter a lot (if he is a con artist or shoplifter) or not at all (if he is a cat burglar). Exposure matters less if others respond by ostracizing the thief than if others respond by beating him up. But whether the other members of the community decide to beat up the thief or ostracize him will depend on their own outside opportunities and how important it is for them to show each other that they are cooperative types.

If it does not matter much—as might be the case during times of tranquillity—people might not punish the thief. A reputation as a cooperative type does not exceed the costs. But sometimes reputations matter a lot. During times of uncertainty and tension, people will punish the thief just to show that they are cooperative types, and as the good types struggle to distinguish themselves from the bad types, the punishment will escalate. This accounts for witch hunts and other extreme, and apparently irrational, communal punishments, which I will discuss in a later section. Everyone punishes the offender (by ostracizing him), despite the triviality of his offense, because his offense serves as opportunity for everyone to signal his reliability.

Shaming sometimes *appears* functional. The group ostracizes the wife-beater, and so husbands refrain from mistreating wives for fear of a similar sanction. This is why some commentators believe that shaming would be a desirable form of punishment. There is no reason, however, to believe that the resulting level of ostracism provides the proper level of deterrence.

A historical example will make this point clear. As mentioned in the last chapter, charivari is a form of communal shaming that has existed in a huge variety of cultures and historical periods, and has disappeared only recently. Charivari has always been directed at sexual misbehavior, such as spousal abuse and quarreling, that might plausibly be thought a form of cheating on a collective endeavor. But charivari has also been directed at simply unusual behavior, such as marriages between people of different ages, interracial or -ethnic or -religious marriages, idiosyncratic sexual practices, and people who hold unpopular opinions. So the fact that a behavior annoys people or threatens public order or manifests shirking may make it salient enough to enable people to draw attention to themselves when they punish the victims; but that is not a necessary condition. The behavior may simply be unusual: that, too, makes it salient. If people join in a charivari to show that they have low discount rates, all that matters is that they incur a cost; it does not matter whether their behavior produces public gains. Their private good (an enhanced reputation) diverges from whatever public good, if any, might be produced through deterrence of certain behaviors.

Mob rule is a similar phenomenon. Although mob action might result in the death of the victim, not his shaming, the logic is the same. It is said that people join mobs because the anonymity of the mob allows them to avoid responsibility for the evil that the mob performs. But what reason is there to suppose that people enjoy committing evil? The reason that people join mobs is that it is better to be a member of a mob than its target. True, fear of the mob might deter people from committing crimes, but there is no reason to believe that the mob provides the proper level of deterrence, especially as the mob's energy seems to be exhausted when blood is spilled, whether or not the victim is guilty (Wyatt-Brown 1982).

This is true because the chief motive for shaming is to enhance reputation, not to do justice. Thus, as long as everyone else is attacking an innocent person, it is better to join in than to risk being thought sympathetic to the mob's enemies. Shaming is also directed at people who are innocent but connected to the offense in some way. Thus, shaming frequently targets the spouse and children of the wrongdoer, as well as the wrongdoer himself. The charivari would often shame the cuckold as much as the adulterer (Wyatt-Brown 1982, p. 450). People have always shamed the illegitimate child in addition to the parents. As we will see in Chapter 8, "punishment" of or discrimination against minorities is a common form of signaling—so the mob will frequently attack members of a minority when a crime is committed, regardless of whether any evidence suggests that they committed the crime.

I am not making an argument about variance. It is true that mob rule will display higher variance than the rule of law enforced by a professional justice system, and therefore risk-averse citizens will prefer the latter, if all else is equal. But my main argument is that even the "average" shaming punishment is not likely to produce the optimal level of deterrence at the margin.

Shaming Penalties Imposed by the Government

Since I have discussed shaming penalties as a means of community enforcement, it might seem odd now to talk about the government imposing shaming penalties. But when I talk of government shaming penalties, I mean punishments imposed by the government in order to facilitate ostracism by the community. To see why this makes sense, distinguish between detection and enforcement. In earlier sections it was assumed that members of the community both detect the deviant (that is, observe his deviant action) and enforce the punishment (for example, ostracize the deviant). These actions imply communication; otherwise, being ostracized does not injure the deviant much nor detection help the community. When the government uses shaming penalties, the two elements of shaming are divided between the government and the community. The government detects and communicates (through the shaming penalty), and members of the community enforce by avoiding the offender and by encouraging each other to avoid the offender.

This division of labor might seem to make sense if the government has a comparative advantage in detecting and communicating deviations, while the community has a comparative advantage in punishment. This is certainly plausible, at least for some kinds of deviant behavior. Perhaps only the government can discover in a cost-effective manner which schoolchildren are carrying guns, because the cheapest way to do this is by putting metal detectors in the doorways of schools, something which community members cannot do through informal cooperation. But the government cannot punish the offenders effectively, because—let us say—schoolchildren are so emotionally traumatized by prison that prison ruins the possibility that they can become productive citizens, and less traumatizing punishments are not practical for the government. Community members can effectively punish gun-toting children by scolding them, throwing them out of their stores and movie theaters, keeping other children away from them, and generally keeping an eye on them. A penalty that publicizes the names of children who carry guns is a government instrument for exploiting independently existing nonlegal sanctions.

A historical example is the pillory. Government agents would capture the offender, verify his offense through trial, and then punish him by placing him in the stocks. The stocks would be located in the public square to maximize the visibility of the offender. Although part of the punishment was the physical pain that resulted from confinement and from being pelted with refuse, an important purpose of the punishment was communicating to other people that the offender had committed a crime. Indeed, in England the authorities occasionally tried to stop the physical abuse by the crowd by ringing the stocks with constables (Beattie 1986). The authorities thus sought to make the punishment more cleanly communicative, rather than corporal.

The pillory is making a comeback. Clients of prostitutes are mostly productive members of society, and they often have families. Putting them in jail deprives family members of their wages and society of their work, and it is possible that these costs exceed any benefits from the marginal deterrence of prostitution. A cheaper and more effective punishment may be the ridicule that friends, colleagues, and family members heap on the client of the prostitute. But if offenders seek the services of prostitutes in anonymous parts of town, they cannot be monitored effectively by the members of their community, and the police and judiciary are necessary for detection. Following arrest and conviction, publication of their names in newspapers aids community enforcement, and may produce a deterrence value superior to fines or imprisonment but at a lower cost. Similar arguments could be made about drunken driving, nonviolent sexual assaults, and various white collar crimes. Many of the people who commit these crimes are middle-class people who care about what other middle-class people think about them.

But an adequate defense of official shaming penalties must overcome the objections described in the prior section. The problem is not violence arising out of mob rule or charivari. Modern governments, unlike their early modern counterparts, can deter such violence by force. The problem is that the government cannot control the level of ostracism that it provokes. I will discuss some historical evidence that the government cannot control the level of ostracism, and, in fact, that the arbitrary severity of shaming punishments led to the abolition of shaming penalties; for now let us imagine a possible consequence of shaming punishments at the present time. Suppose that the government prints in the newspaper or broadcasts on television the names of people convicted of paying for the services of prostitutes. It seems likely that some people will be humiliated by this disclosure while others will not. Why this difference? It must be that unashamed people either belong to a community that does not condemn clients of prostitutes, or have qualities in such demand that they do not have to worry about people ostracizing them (high-status people, like movie actors), or do not rely on cooperative endeavors for most of their gains. But it is also the case that, though the convicted offender might fear humiliation, it may turn out that he receives none. Whether others in fact respond to the exposure of the offender by ostracizing him, or by avoiding him for a few days but then forgetting about the offense, or by doing nothing at all, cannot be foreseen. This is the problem of multiple equilibriums. *Any* amount of punishment can be sustained, suggesting that the actual amount will be sensitive to parameters invisible to a judge.[4]

Morever, it is likely that the victim's spouse will be humiliated by his or her friends, neighbors, and colleagues. As mentioned earlier, participants in a charivari would typically shame the cuckold as well as the adulterer. They would shame the husband as well as the wife who bossed him around, perhaps with the expectation that the prospect of further shaming would deter him from allowing himself to be bossed around further, perhaps with no such expectation. And people would shame the children who are the products of illegitimate union, not just the parents. These historical experiences suggest that if a modern government required, say, the publication of the name of a client of a prostitute, his wife and children would suffer as much humiliation as (more than?) he will.

A response to this argument is that these problems do not distinguish shaming penalties from imprisonment. Prison imposes different levels of disutility on different people. Some people hate being in prison, and other people can tolerate it. Thus, an identical prison term has different effects on different people, just as a single form of shaming has different effects on different people. But this response misses the difficulty. To see why, assume that everyone has identical preferences and opportunities. It follows that a ten-year prison term will injure different people by the same amount. But an identical shaming penalty could still cause different levels of injury. People might pool and ostracize the offender, or people might separate in different groups, some of whom ostracize him and others who do not, or people might pool and not ostracize the offender—all of this depending on random sociological factors such as the general importance of cooperation at a given time for the particular people who witness the shaming, their propensity to gossip with others, and the availability of alternative signals for any one of them.

Although imprisonment may produce a similarly variable stigma, that is not the sole purpose of imprisonment; to the extent that it is the purpose, or is made the purpose, then imprisonment is just another shaming penalty. It is better to think of the stigmatizing effect of imprisonment as an unavoidable side effect that could be removed only at great cost—though the government currently does relatively little to spread this information and often restricts it.

Government Shaming as Information Revelation

There is another way of looking at government shaming. Putting aside the moralistic cast of shaming, one can think of it as a form of information revelation. After the government tries, convicts, and releases (if imprisoned) an offender, it has generated valuable information about the offender. This information is put on his record. Governments face an important choice of whether to release an offender's record to the public. At one extreme, the government could expunge the record. Governments do expunge criminal records quite often, not just for juveniles but also for adults who have been convicted of certain relatively minor crimes (mostly, drug crimes) and who have engaged in good behavior over a period of time. Governments might also maintain the record for law-enforcement purposes but refuse to release it to interested private citizens, such as potential employers of the offender. Some governments do release criminal records to potential employers, or do so if their employees have sensitive tasks (for example, daycare). The federal government restricts the ability of credit reporting agencies to report past arrests and convictions in credit reports. At the opposite extreme, the government releases a criminal record to anyone who asks for it, or even publicizes the offender's record, or parts of it, as required by Megan's Law. The fact that governments often conceal criminal records is powerful evidence that governments worry about the stigmatizing effects of criminal records; but the variety of responses suggests that the theoretical case against shaming penalties is not airtight, that in fact their suitability is an empirical matter.[5]

Government Shaming Penalties Imposed on Members of a Hostile Subcommunity

The arbitrariness of the sanction is not the only objection to the official use of shaming penalties. To analyze a different one, I will put this objection aside, and assume, for the sake of expository clarity, that shaming penalties do not produce such arbitrary results. The commitment model described in Chapter 2, which should not be confused with the signaling model, suggests further problems with government shaming.

Imagine that society can be divided into Xs and Ys, and X and Y refer to people with different physical characteristics or other characteristics that can easily be observed or discovered, such as race, ethnicity, immigrant status, or religious practices. Within each group, there are good and bad types. Assume that the Xs are wealthier and more politically powerful than the Ys—perhaps because they are more numerous, or because they are descended from a politically powerful group. Under such conditions, it is possible that members of X will signal their cooperative propensity by discriminating against the Ys. Discrimination functions as a signal because it is costly (one gives up the opportunity to do business or socialize with Ys who are good types), and because conspiracy theories based on traditional fears of people with physical or cultural differences, or people who are recent arrivals, provide a focal point: given these beliefs, those who discriminate against outsiders will appear to be loyal to the insiders. In equilibrium all or most Xs will avoid Ys (and those who do not will be ostracized by the Xs and may seek refuge with the Ys), and all or most Ys will avoid the Xs, falling back on each other when it is desirable to enter cooperative ventures. Finally, we assume that because of their political and economic dominance, members of X control the government, including the criminal justice system.

Note that X may or may not be a numerically larger group. In discussing many areas of the United States, and especially the past, it would be accurate to say that the Xs represent whites and the Ys represent blacks or members of other racial or ethnic groups, and Xs are more numerous than Ys. But the Xs might consist of the ruling class, a tiny minority of people with aristocratic lineage or great wealth, while the Ys consist of the masses.

If the Xs operate the criminal justice system to the disadvantage of the Ys, and if, in general, the Xs have dominant political and economic power and use it in a discriminatory way, the Ys must rely heavily on each other for protection against criminals. They form a solidary group (see Chapter 2). Because of the danger of free-riding, Ys signal their loyalty to the group. One powerful signal is to engage in activities that the Xs disapprove of and even punish. The Y who engages in such activities is a more reliable cooperative partner than other Ys are, since he has cut off some of his outside opportunities (those involving Xs) should he be ostracized by Ys for cheating.[6]

Criminal punishments will have different effects on Xs and Ys. Because Ys obtain reputational payoffs from other Ys when they become subject to criminal punishments, a criminal punishment will deter a Y less than it will deter a similarly situated X, who obtains no reputational payoff. A Y who has a criminal record will not be trusted by Xs, but because this Y must therefore rely on Ys for cooperation, this Y is less likely to cheat a Y than other Ys are. Criminal punishment might for this reason even encourage Ys to commit crimes. This hypothesis is supported by data that show that increasing the punishment for certain crimes deters criminal activity by members of the in-group but encourages criminal activity by members of the out-group (Sherman 1993).

I have been speaking of criminal penalties generally, not just shaming punishments. The communicative element in shaming punishments, however, is likely to heighten the differential impact on Xs and Ys. In particular, shaming penalties imposed by the government on members of a hostile subcommunity are likely to encourage deviant behavior, because the shaming penalty communicates to members of the dominant community that the offender is a deviant, thus causing them to avoid him, and thus enhancing the offender's reliability when cooperating with members of the subcommunity. As noted before, Y has an incentive to get himself punished by the Xs, or to commit a crime and thus make himself liable to punishment by the Xs—even if he does not have a particular desire to commit a crime—because this semi-voluntary punishment is a reliable signal of his loyalty to the other Ys. When the punishment is communicated by the government, the signal is much more focused and readily understood by other Ys. A criminal punishment that was kept hidden from the public would deter a Y from crime more effectively than a criminal punishment that broadcasts his offense.

To be sure, some or many Ys may also take the punishment as a signal that the offender is a low type—that is, a person who engages in crime because he is impulsive or unconcerned about the future. If the person is a low type, then the Ys will ostracize him. But the less the Ys trust the government's motives—the more they believe that the criminal justice system is infected with a political agenda—the more likely it is that they will interpret a punishment by the government as a signal that the offender is loyal to the Ys. In such circumstances, the government should punish offending Ys as quietly as possible, and avoid shaming penalties at all costs. This may explain why, despite the increasing popularity of shaming penalties, including the ones described above, for crimes frequently committed by members of the middle class, there has been less enthusiasm for using shaming penalties for crimes frequently committed by blacks and other ethnic minorities.

The status of minority prisoners in some inner city communities is similar to the status of political prisoners. The difference between an ordinary prisoner and a political prisoner is that the latter gives up his freedom in order to aid a subcommunity that has hostile relations with the government. Thus, a political prisoner has high status in the subcommunity and among the subcommunity's admirers. To be sure, a person who commits a crime like shoplifting and then claims that he did so to help the subcommunity may not be believed. The most credible political prisoners are those whose crimes do not benefit themselves: they violate laws against association in an effort to form political parties, or they rob banks but lead Spartan lives and use the proceeds of bank robberies to support a revolutionary organization. They engage in disruptive behavior that has no evident benefit for themselves, like occupying streets and businesses without stealing property or engaging in violence. Risking one's freedom is a credible signal that one is loyal to the community only when it is not seen as a risk rationally taken for the sake of private gains. Similarly, gang members who provide valued services to a community, like protection against criminals and members of rival gangs (Jankowski 1991, pp. 180–93), will be honored, not condemned, when they are carted off to prison by a hostile and unresponsive police force.

There is one more complication: even if the Ys suspect that the offender is a genuine deviant, and not a loyal member of their subcommunity, the logic of signaling may prevent them from acknowledging this. As long as there is some doubt regarding the matter, people will signal their loyalty to the Ys by protesting the innocence of the offender. Some people may genuinely believe this; others will agree with the emerging view just to show that they are loyal to the subcommunity. Disagreeing with this view will be taken as a signal that one rejects the subcommunity. This will strain relations with members of the dominant community, for whom the deviance of the criminal is obvious and who thus cannot see why members of the subcommunity will not acknowledge it. Famous recent examples include O. J. Simpson, Marion Barry, and Tawana Brawley. Among some blacks, refusal to concede the guilt of these offenders serves as a signal of their loyalty to the subcommunity. But this is a general phenomenon, true of all ethnic minorities and other groups for which intra-group loyalty is crucial because of the hostility of dominant groups.[7]

In American cities police forces have historically been dominated by the white majority. After some integration, blacks continue to believe plausibly that the criminal justice system treats whites better than blacks, and that whites continue to discriminate against blacks. Committing a crime can serve as a commitment mechanism, when the relevant community is an inner-city neighborhood that is suspicious of the white establishment and fearful that some of its members are disloyal (for example, they work for the police or seek to escape the community). And, indeed, in inner city black and Hispanic communities imprisonment and other punishments may confer status on offenders, at least among gang members.[8] White officials believe that if more blacks are involved in criminal justice institutions, it is less likely that black communities will regard punishments as racially or politically motivated (Jankowski 1991, pp. 260–63). However, this strategy faces the problem that blacks might regard black participants in the civil justice system as traitors to their community or that blacks who participate in the system will refuse to follow its rules. Hence the fate of Clarence Thomas and Christopher Darden, to varying degrees, on the one hand; and the dramatic phenomenon of jury nullification where black jurors refuse to vote to convict black offenders, on the other. Although no doubt integration works more effectively in some contexts than in others (the most promising examples occur when already acknowledged leaders from the black community obtain visible positions of power in the system), it will not by itself solve the problem that punishment can confer status.

The Creation of Deviant Subcommunities

In discussing shaming penalties I have assumed that a hostile subcommunity already exists, and then have asked what effect shaming penalties might have on the behavior of members of that subcommunity. But hostile subcommunities do not arise by themselves; they are made. One way they are made is, as Braithwaite (1989) has observed, through the criminal law.

To see how this works, recall that signals are necessary for the creation of relationships and communities but that they are historically arbitrary. Typically, an action that serves as a signal of loyalty to the community imposes different costs on different people. This means that while the average person rationally incurs the cost of the signal in order to obtain the gains of cooperation, people for whom the action is very costly do not. These people will then be ostracized by members of the dominant community. They respond by forming subcommunities.

An example is ideology or religious belief. If a community has certain theological commitments that I share, it is costless for me to signal my loyalty to it. If I do not care much about theology, it is costly, but maybe less costly than forgoing gains from cooperation. If I care deeply about my religion and my beliefs contradict those of the community, I may incur greater costs (in terms of psychological or spiritual well-being) than I gain (materially or psychologically or spiritually) from membership in the community. Since it is too costly to convert, I seek out like-minded people to form a subcommunity that distinguishes itself from the dominant community. Or consider the creation of subcommunities in response to racial or ethnic discrimination. I cannot change physical characteristics of my body, or my ancestry, so if everyone discriminates against me on the basis of these characteristics as a way of showing each other that they are loyal to each other, then I cannot join their community. I am forced to form a subcommunity with people who share my characteristics.

Notice that in both cases, I now have my own cooperation problem, for I must show people in the excluded group that I am loyal to them. As Chapter 2 discussed, a reliable signal in a subcommunity is behavior that contradicts the signals that are used in the dominant community. Signals that offend members of the larger group are especially effective at maintaining loyalty in the subcommunity, because those in the subcommunity reduce the chance that they could be accepted by people in the larger group. When a teenager makes himself obnoxious to adults or to other teenagers, he reduces the value of his opportunities outside his circle of friends. Since he therefore loses more if he cheats, is detected, and is rejected from his group, the other group members can trust him more than before. It is one more step to argue that criminal behavior directed against the larger group can be a signal of loyalty to the subcommunity.[9]

As a dissolute youth, St. Augustine stole some pears with his friends, then wasted them by feeding them to hogs. The wastefulness of the activity is supposed to be shocking; we would have felt more sympathy if the youths stole the pears because they were hungry and wanted to eat the pears. But all signals are wasteful; and the theft and pointless destruction of pears were signals that Augustine and his friends sent to each other. As he notes, he would never have done this if he had not been with his friends (Augustine 1961).

The signals that happen to prevail in the dominant community will often determine the kind of deviant community that will arise. When the signal consists of sexual self-restraint (whether with respect to frequency of sex, number of partners, kinds of sexual behavior, and so on), those who for whom such restraint is especially costly will form the subcommunity. When the signal consists of discrimination against members of a group with identifiable physical characteristics, those people and their friends and admirers will form the subcommunity. When the signal consists of repression of various forms of hedonistic activity (drugs, alcohol), those who want to form a subcommunity will engage in that activity. When the signal consists of the public expression of certain ideological commitments, the subcommunity will reject those commitments. Thus, an action and its opposite can both be signals.

To illustrate this point, imagine that at time 0 some people buy sex from prostitutes without incurring much shame. At time 1 the government decides to punish this behavior. It can choose (1) a prison term that harms offenders by an expected amount that exactly offsets the average utility that people obtain from committing the offense; or (2) a shaming penalty that produces the same amount of harm. (I am assuming that the government can control the extent of shaming.) In both cases, the people who obtain above-average utility from the offense will continue engaging in it, while all the others will stop. Thus, the deterrent effect of the two punishments is (by construction) the same with respect to immediate incentives to engage in the behavior. However, if the shaming penalty effectively communicates the offenders' behavior, and the dominant community deeply disapproves of it, people will respond by ostracizing the offenders. Having lost legitimate opportunities for gain, the offenders must turn to a life of crime, and if they are able to cooperate with each other, they will form a deviant and hostile subcommunity that poses a threat to the dominant community. In contrast, the undeterred offenders who are imprisoned and released under the first system have no incentive to sacrifice the benefits of a legitimate occupation for the dubious gains of membership in a deviant subcommunity.

An ordinary criminal punishment, like imprisonment or a fine, may have effects that are similar to those of a self-conscious shaming penalty. People do stigmatize ex-convicts; employers check for criminal records. But the question is whether the government should enhance the reputational effects of imprisonment and other punishments by publicizing the names of offenders or by shaming them in other ways, or should not. The reason the government should not normally enhance these reputational effects is that doing so could drive people to form deviant and hostile subcommunities.

Some Historical Evidence

The historical decline of shaming punishments includes (1) the elimination in advanced countries of public executions, and especially of hideous public executions; (2) the elimination of flogging, the stocks, special clothing, and other forms of public humiliation; and (3) the elimination of brands, tattoos, amputations, and other disfiguring punishments that identify an ex-convict to anyone who meets him. A focus on these trends helps to substantiate the theoretical arguments.

One would like to say that the elimination of these punishments is due to the improvement of morality. It is just as likely, however, that the causation is the reverse: people cannot tolerate public torture and execution because they are no longer used to observing them—just as people can no longer watch the slaughtering of animals because they no longer live on farms. If this is the case, the source of these developments in criminal punishment must lie elsewhere.

Public executions. Public executions humiliated the offender as well as killing him. Contemporary observers made much of the terrified and wretched figure of the condemned as he was led out into the gaze of the crowd. Usually there would be a carnival atmosphere. Merchants would hawk souvenirs, children would play among the legs of the observers, there would be games and songs. Frequently the crowds would be moved by pity for the condemned, or by outrage. But in either event disorder often resulted (see Gatrell 1994, pp. 90–105, 266–72, 589–91; also Beattie 1986). Whether the danger posed by the crowds outweighed the deterrent effects that would otherwise result from the publicity of the execution is a hard question, and certainly early on the authorities believed that public executions had a deterrent effect. But by the nineteenth century it had become widely acknowledged that alternative forms of punishment, such as private execution and imprisonment, would produce less disorder, and accordingly public execution was abolished in England in 1868.

Although the Xs in, say, early modern Europe consisted of tiny elites, rather than racial or ethnic majorities, they were no less feared, despised, and mistrusted by the masses. Thus, it was always an open question whether someone hauled up on the scaffold was a threat to the poor or a savior, and so the communicative effect of public shaming could just as easily have been to amplify the government's illegitimacy as to heighten the penalty for deviance. The condemned might be seen as someone who had sacrificed himself for the good of the people, and his punishment seen as a vindication of his commitment at the expense of the legitimacy of the government. The most dramatic instance of this is martyrdom, where the execution of the martyr casts doubt on the government's motives rather than on the martyr's cause. Over the long term the martyrdom of Christian believers enhanced the legitimacy of Christianity and prepared the way for its establishment as the official religion of Rome.

Flogging and pillories. Public punishments short of death communicated the offender's crime to the watching crowd, and via gossip, to the entire community (Beattie 1986, p. 464). The pillory is the purest example of this phenomenon. The offender is literally held out for everyone to see; there is no corporal punishment administered by government officials. Although the crowd often threw objects at the offender or beat him up, this was not the point of the punishment; as noted before, government officials would sometimes try to protect the offender from the crowd (Beattie 1986, p. 616).

The problem with pillories is that the severity of the punishment was unpredictable (Beattie 1986, pp. 466–68). Sometimes (though rarely), the crowd would release the offender, as in a few cases of sedition against an unpopular government. Sometimes, the crowd would kill the offender by tearing him from the stocks and beating him to death or by hurling objects at him until he died. Even when the crowd merely mocked the offender, the reputational effect of the pillory could not be gauged in advance. It must always have depended, in part, just on who showed up and saw the humiliation of the offender, how eager they were to spread the news, and what other things they had to do with their time. Fitting the punishment to the crime was an extremely crude business. Indeed, judges sentenced offenders to the pillory with great frequency only to avoid the extreme alternatives—death, transport, and similarly fatal or near-fatal punishments. Reformers lobbied for imprisonment because it would allow judges, by the setting of different prison terms, to have greater control over the severity of punishment (Beattie 1986, p. 608).

One might speculate that even in the case of flogging, the point of the punishment was not to cause pain but to cause public humiliation. The pain was not for the benefit of the victim, but for the benefit of the onlookers. This theory—that the point of corporal punishment is not to cause pain to the offender but to imprint an image in the observers' memories—is supported by the practice of administering narcotics to the victims of grisly executions, and the use of powder to hasten the conclusion of the auto-da-fé. If the purpose of such unpleasant executions was to cause pain, why take steps to reduce it?

Flogging and grisly executions can be interpreted as a form of pillorying where the offender is beaten or killed or mutilated not because a painful death would deter people from crime more effectively than a painless death, but because people are more likely to *remember* the identity of the offender (in the case of flogging) or the punishment of an offense (in the case of execution) if the punishment is unusual, severe, or disgusting. Thus did one man ask "the Surrey magistrates to alter his sentence from a public to a private whipping so that he would be able to continue to find employment in Southwark and to continue to support his family." (Beattie 1986, p. 464.) I will pursue this theme below.

Branding and other markings. In the early modern period in England a person accused of most serious crimes could plead "benefit of clergy," that is, that he belonged to the clergy and was not subject to the crown's jurisdiction. Because of the poor state of records, widespread illiteracy, and extreme ambiguity about what the clergy was, courts could not usually verify whether a person belonged to the clergy simply by consulting documents or asking a church official. Accordingly, (almost) anyone could plead benefit of clergy, but, under the law, one could plead benefit of clergy only once. The next time one was accused of a crime, the defense was no longer available. To keep track of who had pled benefit of clergy and who had not without benefit of record-keeping, the courts required that those who plead benefit of clergy be branded on the thumb. Thus, if a person who pled benefit of clergy already had this brand, the defense would not be allowed.

Over time the authorities realized that the brand on the thumb was a poor means for recording one's use of benefit of clergy. The brand was often badly applied and the skin healed. The authorities responded to this problem by enacting a statute that declared that those who plead benefit of clergy would be branded on the cheek. But this method, too, was ultimately abandoned (with a return to branding on the thumb). "The reason given in the new statute [for abandonment of branding on the cheek] was that the permanent and visible stigmatization of prisoners had made them 'unfit to be intrusted in any service or employment to get their livelihood in any honest and lawful way,' and by thus making them 'more desperate' had obliged many to follow a life of crime." (Beattie 1986, p. 464.) Because employers avoided hiring people who were branded on the cheek, no doubt on the basis of the inference that those who had committed crimes would be dishonest employees, these people could not earn a living honestly. So they had to join gangs and earn their living through crime. Here, although the intent was not, as far as I can find, to shame those who pled benefit of clergy, but just to record their use of that defense, the brand nevertheless had the effect of a shaming penalty. The authorities abandoned branding on the cheek, presumably because they believed that underdeterrence of crime could be tolerated when the alternative was the creation of criminal gangs that posed a significant and persistent threat to social order.

History reveals two problems with shaming punishments. First, these punishments are messy. They are intended to exploit the independent force of crowd dynamics, but crowd dynamics are unpredictable. A punishment whose severity is unpredictable cannot be used by judges to achieve marginal deterrence. Second, these punishments created deviant subcommunities. When it is very easy for people to identify past offenders, they will avoid them, so the offenders are driven to join criminal gangs. Both of these concerns lay behind the call for reform in the nineteenth century (Beattie 1986, pp. 500, 614, 616).

Both concerns have continuing relevance. The use of shaming penalties against drunken drivers, smokers, and the clients of prostitutes are likely to have perverse results. The punishment or the activity might be seen, especially in deviant subcommunities, as honorable, a signal of commitment to the subcommunity. Consider the current crackdown on cigarette smoking: this has heightened the association of cigarette smoking with nonconformity and rebelliousness. A skilled movie maker can with great economy convey the independence of a character by placing a cigarette between his lips, increasing the allure of cigarettes to youthful rebels. Cigarette smoking among teenagers increased by more than 30 percent in the last five years, as did the frequency of smoking in movies, even as smoking among adults declined (Klein 1997). But I do not claim that the movies are causing teenagers to smoke. Both movie makers and teenagers are exploiting the increasing social and political intolerance of smoking for their own purposes.[10]

Redemption, Mercy, and Rehabilitation

Reformers in nineteenth century England sought to eliminate shaming penalties under the slogan of rehabilitation, according to which the goal of criminal punishment is to transform the offender into a good citizen. Rehabilitation is a secularized version of two important theological ideas: redemption and mercy. Redemption and, to a lesser extent, mercy refer to the forgiveness of a person who has committed a crime against a community. Although the punishment may be waived or reduced, redemption is not the same thing as a declaration that a person is innocent of a charge. What it means is that the person has committed the crime, and deserves expulsion from the community, but he is now accepted back into the embrace of the community.

Redemption and mercy play a powerful role in Christian theology, but we see these values in secular law as well. We have seen examples of expungement of criminal records.[11] In some countries a person who commits a crime that is later expunged from his record has a claim in defamation against anyone who subsequently mentions the crime publicly.

These laws cannot be easily placed in the framework of the classical economic theory of crime. Information about past crimes is relevant to employers and others who deal with former offenders. If the stigma creates too great a penalty, resulting in overdeterrence, then the logical response is not to prohibit the spread of information but to reduce the original punishment. The rational criminal will account for forgiveness by calculating a smaller expected punishment.[12]

An explanation of these laws rests on the unpredictable effect of information transmission. Although imprisonment imposes a relatively predictable cost on the offender, the reputational effect of imprisonment is hard to predict. The state might not care that some offenders receive this additional unpredictable reputational loss while not wanting all offenders to receive this loss. But which offenders might fall into this latter category? To answer this question, I will divide criminals into four classes: (1) those with intrinsically anti-social tastes, like those of the sadist; (2) those who had extremely bad outside (that is, legal) opportunities at the time they committed the crime, like the starving person who steals a loaf of bread; (3) those with a high discount rate, or, what would normally be called impulsiveness; and (4) those who commit crimes to demonstrate commitment to a deviant subcommunity.

The criminal law distinguishes people on the basis of which motivation predominates. The sadistic criminal is not forgiven; the punishment is not mitigated. Overdeterrence is not a concern. The person who commits a crime because of bad opportunities will not commit a crime if his opportunities improve; it is at this person that prison education and training are aimed. The person with a high discount rate might later develop a low discount rate as he matures: hence expungement of juvenile records and of the records of people who have refrained from criminal activity for a period of years. Finally, the person who begs for mercy, who forswears further association with the deviant subcommunity, in the process commits himself not to rejoin it. Self-abasement, implied by the phrase "*beg* for mercy," is critical: it is another costly action that can serve as a signal to all concerned that the offender switches allegiance from the deviant subcommunity to the dominant community.[13] To be sure, the remaining members of the subcommunity might not regard this signal as authentic, though they will have to if the offender demonstrates his commitment further by informing on his former colleagues.

The expungement of criminal records, the refusal to publicly humiliate people, the abandonment of branding, tattooing, and public executions—these are all closely related. All of these restrictions on the publication of a person's crimes are based on the fear that people who are stigmatized will opt out of legal society where they pose a greater threat than they would otherwise.[14]

But if this is so, why should criminal records be maintained at all? The answer is that the information is valuable if used properly. The pedophile should not become a schoolteacher; the robber should not become a security guard. The police should be able to narrow down a list of possible suspects by looking at their records. The use of information in this limited way, however, is nothing like shaming; and the offender can protect himself from exposure by declining to apply for sensitive jobs.

Why Are Imprisonment and Corporal Punishment Shameful?

An issue I have treated only in passing is worth focusing on: Why are some punishments more shameful than others? Kahan (1997) says that imprisonment is shameful, whereas paying a fine is not shameful or is less shameful. Why should this be the case? This question is important: if imprisonment is shameful only if (for example) it is uncommon, throwing lots of people into prison in order to shame them might reduce the shameful effect of being a prisoner.

Why is imprisonment more shameful than being fined? Why can't the government say: "People who are fined are much worse people than those who are imprisoned, or are just as bad"? One might argue that it is shameful in a free society to be deprived of one's liberty, or to be physically constrained. But no one would stigmatize a person who has been quarantined because he or she has traveled from a country where an infectious disease is widespread; yet quarantined people are effectively imprisoned for the length of the quarantine.

The shamefulness of the punishment for a crime is a function only of its communicative effectiveness. The communicative effectiveness of a punishment is a function of (1) its visibility, (2) its memorability; and (3) the strength of the association between the punishment and the badness of the people who are punished.

Visibility. One explanation of the relative shamefulness of imprisonment is that imprisonment is more visible than fining. A person who is imprisoned is absent from family and workplace, so people are constantly reminded of the person's wrongdoing by his or her absence. A person who is fined is not visibly punished so long as this person's checkbook is concealed from the public. To be sure, a sufficiently high fine will result in a decline in a standard of living, and this may be as visible as imprisonment is. Maybe a person could attribute the decline to some other factor ("I had to help out a sick relative"), but as long as one can lie credibly, one can explain away long absences as well ("I was visiting a sick relative," "I was away on business").

Memorability. A lot of people commit crimes and receive punishments. Citizens thus tend to forget that a person was punished or why that person was punished. As mentioned before, a theory of the grisliness of executions in the past is that people remember a grisly execution more vividly than an ordinary execution. People will also remember a flogging more vividly than a pillorying. The crack of the whip, the screams for mercy, the smell of blood—these are not likely to be forgotten even in a brutal society. In our less brutal society, many people pay fines; relatively few go to prison; so people are more likely to remember that an acquaintance served a prison term than that he paid a fine.

Badness. A punishment causes shame only if the person being punished is subsequently avoided or ostracized, rather than celebrated. In other words, ordinary people must infer from the fact that a person has been punished that this person is bad. If the government threw only good people in prison, then imprisonment would not be stigmatizing.

People might not know whether an offender really committed a crime or whether his or her crime was really harmful to society. This raises the possibility that a punishment is shameful only if the government can convince the public that the offender engaged in bad behavior. Imprisonment is more shameful than a fine because imprisonment is a better signal of the government's attitude toward the convict than fining is. Imprisonment of a person is extremely costly; fining a person is not nearly as costly. Thus, by imprisoning a person the government credibly reveals that it considers that person dangerous to the community (or itself); fining is not a credible signal. Indeed, concern that a corrupt government uses criminal fines to enrich itself is a common theme in history, a recent manifestation of which is the charge that the police target wealthy drug users in order to enrich themselves under the forfeiture laws. The use of prisons, rather than fines, is a way for the government to show that it acts in the public's interest, rather than in the interest of government officials. (A possible theory for why the death penalty process is more expensive for the government than life imprisonment is that this is the only way for the government to show that the death penalty is legitimate.)

This theory is consistent with the observation that imprisonment enhances status when the convict is a political prisoner. Imprisonment allows the political prisoner to claim that he or she is really a danger to the government. But this raises problems for the government. By imprisoning a person or even by fining him or her, the government reveals that it considers that person to be a threat, but in doing so it also reveals that it (the government) is vulnerable. If the government reveals that it is vulnerable, and many citizens oppose the government, then they will become more confident about their ability to overthrow the government, and act on this increased confidence. The idea that a tolerant government may be more stable and powerful than a government that punishes opponents has a paradoxical ring, but it also has deep roots in political theory.

The discussion also brings out the problem with assuming that imprisonment is intrinsically more shaming than fining is. Suppose the government required people convicted of trivial offenses, like littering, to spend one hour in jail. It is unlikely that an offender of this kind would be stigmatized, just because he has spent some time in jail. By the same token, if the government only threw its political opponents in jail, then imprisonment would not be shameful. *The reason that imprisonment is stigmatizing is that only (or mostly) bad people spend time in prison.* Ordinary people therefore infer from a person's having been in prison that this person is bad and should be avoided. If too many people are put in prison who are not considered bad (drug users? white collar criminals?), imprisonment will lose some of its stigma. Even a thief could be honored if he was imprisoned in the Bastille. As Emerson said of John Brown, he "will make the gallows glorious like the cross" (Furnas 1959, p. 381). People object to community service as a criminal punishment, because if criminals are forced to clean the street, then the status of street cleaner will decline. Ordinary people will infer that an honestly employed street cleaner is bad if most street cleaners are criminals who are being punished (Kahan 1996). It seems likely that in an earlier time, hanging was shameful while decapitation was not shameful, just because commoners were hanged and aristocrats were decapitated. It is hard to know what started this pattern, but once it was started, hanging a nobleman denied him and his family their aristocratic pretensions.

A final point is that if the stigma associated with imprisonment is desirable because of its enhancement of the deterrent effect, it is possible that imprisonment of too many people will reduce deterrence. When the prison population rises beyond a certain threshold, people will stop shunning former prisoners, because they form such a large proportion of the population. If one avoided all former prisoners, one would not be able to afford to hire any employees. Because former prisoners are not just very bad people, but also not-so-bad people, employers and others will not infer that a former prisoner is necessarily a bad person.[15]

The Deterrence Model and the Normative Model

Much is made in criminology of a distinction between the deterrence model and a so-called normative model (Tyler 1990). The former is the economic model: people commit crimes unless the law deters them with adequate punishments. The latter model holds that people commit crimes unless the criminal law has "legitimacy"—that is, either people believe that the law is morally right and they seek to engage in moral behavior, or people believe that they should follow any law that is created by a legitimate authority, and they believe that the current authority (the government) is legitimate (that is, the officials act in the public's interest, or were properly elected, and so on). Not surprisingly, the evidence suggests that both models only partly explain people's behavior (Tyler 1990).

The two approaches can be unified using the approach I have been discussing. People signal to each other that they are cooperators by engaging in costly actions. The normative model is thus interpreted to mean that one signal of cooperativeness is obedience to a plausibly desirable law *even though the gains from violation exceed the expected cost of legal punishment.* The tension between the normative model and the deterrence model is apparent. If I decline to litter because I fear that I will be arrested and thrown in jail, the deterrence model explains my action. If I decline to litter even though the littering law is not enforced, then the normative model explains my action. My suppression of the impulse to litter is a signal to others that I am a cooperator. But my failure to litter will be a signal of cooperativeness only if most people agree that littering is bad behavior—if, in other words, the law has legitimacy. So if a law has legitimacy, it is more likely that people will obey it not because they derive utility directly from obeying a legitimate law, but because they are likely to obtain future returns when others see them obeying a legitimate law.

7

Voting, Political Participation, and Symbolic Behavior

An important but neglected aspect of the law is its symbolic function. A symbol is an image that refers to a system of beliefs that are generally known, if not necessarily shared, by the person who observes the symbol. Symbolic behavior is an agent's use of a symbol to show that he shares or rejects these beliefs, and usually involves either an act of respect for the symbol or an act of rejection. When I wave an American flag at a parade, for example, I do not do so because I enjoy the physical exercise, the way I might enjoy hitting a punching bag. I do so in order to persuade people that I have certain beliefs about the United States. When I burn or mutilate a flag, I draw attention to my rejection of these beliefs.

It hardly needs to be said that symbolism is important in politics and the law. Flag desecration itself riles emotions periodically, most recently in the early 1990s, but also during the Vietnam War and as far back as the end of the nineteenth century (Goldstein 1996). The most recent eruption resulted in state legislation, a significant Supreme Court opinion, and even efforts to amend the U.S. Constitution. And yet what is at stake? In many countries, the flag means nothing, flag desecration is unknown, and citizens fight over other symbols, like the lyrics of the national anthem or the treatment of the national bird.

Other areas of American law concern symbols. The jurisprudence on the religion clauses of the U.S. Constitution concerns the use of public property to display religious symbols. Voting is essentially a symbolic act, since an individual's vote has virtually no influence on the outcome of elections, so laws that regulate voting are laws that regulate symbolic behavior. Controversies over national holidays (like Martin Luther King, Jr. Day), whether there should be a national language, apologies to historically victimized groups, and so on, are essentially controversies about symbolic behavior. Yet, despite the importance of symbolism in politics and the law, the legal literature lacks a methodology for understanding it.[1]

A Model of Symbolic Behavior

I will limit myself to a discussion of patriotic symbolism, by which I mean symbolic behavior that shows one's commitment to the state, but the discussion can be extended to symbolic behavior that shows one's commitment to a religious group or to an ethnic or racial group (see the next chapter). The model of patriotism is a simple reinterpretation of the cooperation model described in Chapter 2.

Patriotic behavior can be a signal that one belongs to the good type. The reason is that typical patriotic behavior—volunteering for the army, joining rallies and parades, waving the flag, voting—does not satisfy preferences that most ordinary people have, as everyone recognizes. No one displays the American flag outside his or her house on July 4 for the sheer aesthetic joy of it. If one really thought that the flag was uniquely beautiful and enhanced the appearance of one's house, one would display the flag every day, not just on July 4. This is not to say that patriotic behavior is always a clear signal. In the past the army attracted adventurers and today it supplies valuable training. But when people join the army in a time of war, most expose themselves to costs that exceed whatever gains might be expected. Similarly, joining a parade might be fun independently of its patriotic meanings; but to join a parade during, say, the Vietnam War, or during Veteran's Day in the pouring rain, is not intrinsically enjoyable for most people. In this way, patriotic behavior is like gift-giving, where there is always some ambiguity whether the donor acts from altruism (in which case the signal is costless and therefore meaningless) or from a desire to reveal type. "Symbolic behavior" is thus simply signaling behavior. Patriotic symbolism is signaling behavior that exploits people's prior beliefs that actions relating to the good of the country are generally costly actions.

A stylized example can be used to illustrate these points. The McCarthy era began after the explosion of a hydrogen bomb by the Soviet Union and the exposure of Soviet spies in the American government, events that heightened fears about the security of the United States and provoked concern about national cohesiveness. Although people always worry about the trustworthiness of their social and business partners, increasing international tensions heighten these worries. The victim of a partner's opportunism incurs a greater loss, if everything else is equal, in a wartime economy characterized by scarcity than in times of peace and robust economic growth. In an atmosphere of crisis, traditional signals of patriotism, such as attendance at parades, are likely to be insufficient to distinguish good types and bad types. As the cost and likelihood of being shunned increases, bad types will find it worthwhile to invest in the relatively inexpensive signals. To distinguish themselves from the bad types, then, good types must discover more expensive signals. The signal proposed by McCarthy was the shunning of communists and anyone else who expressed skepticism about traditional American political institutions and values. In certain areas of the American economy, such as the entertainment industry, the proposal was dramatically successful. To avoid losing their jobs, people informed on or shunned members of the Communist Party. At the time, this behavior was viewed by many people (though not, of course, by critics) as patriotic behavior, similar to volunteering for the army and civil defense and (to take examples from World War II) recycling rubber and tending victory gardens, and to attending parades and memorials, observing national holidays, and showing respect for the flag.

The significance of world events was twofold. First, they enhanced the felt need to avoid bad types. In times of peace, entering a relationship with a bad type may not be so dangerous: the bad type may cheat you, but little is at stake anyway. In times of war, entering a relationship with a bad type is dangerous; because resources are pinched, and more is at stake. One might object that while this may be true during real emergencies, when goods are scarce and ordinary conveniences disappear, a buyer of widgets is not going to lose confidence with the seller just because of international tensions. This is a reasonable objection. Second, world events supplied the focal point of the signaling equilibrium. If the Soviet Union is the threat, and the Soviet Union is controlled by communists, then, the reasoning goes, maybe American communists are also a threat. Shunning American communists and their allies, as long as it is costly, reveals one's commitment to cooperation with non-communist citizens. The buyer of widgets and the seller of widgets reaffirm their trust by condemning communists, not by praising Stalin.

McCarthy, a classic norm entrepreneur, did not fabricate the association between communism and subversion, but he strengthened the association by drawing attention to it. Before the McCarthy era, avoiding communists was not a powerful signal of loyalty. During the McCarthy era, it became such a powerful signal of loyalty that a separating equilibrium was created. Only after some years did this equilibrium collapse, when Americans came to believe that the Soviet threat was exaggerated, or that the threat from bad types was exaggerated, or that the cost of erroneously shunning people who were not bad types was too high.

The last point is important. The patriotic signal would not have been issued by good types who cared deeply about the good of the country and who objected to McCarthy's tactics or were committed to communism, and yet these people would have been ostracized. There is a difference between one's type—which refers to one's discount rate—and one's beliefs about what is best for one's country. The demise of McCarthyism resulted in part from the realization that good types were being shunned along with the bad. Once people stopped believing that only good types supported McCarthy and only bad types opposed McCarthy, support for McCarthy became a costly action that failed to reveal that one belonged to the good type. The separating equilibrium collapsed into a pooling equilibrium in which no one (or few people) sent patriotic messages and no one (or few people) inferred from the failure to send such messages that a particular person would be an unreliable partner.

The model suggests that when tensions increase, people will signal more. The point is not just that during wars, plagues, civil unrest, and massive social change, people will care more about their reputation, because they must trust each other more, and can rely less on the state. What is special about signaling is that as it becomes more important to send the right signal, behavior departs more and more from the sort of ordinary value-maximizing behavior in which people engage when they cooperate for the purpose of producing and sharing a surplus. Behavior becomes divorced from value-maximizing behavior, and so it appears strange to those looking back at it from later periods. Behavior becomes excessive, intense beyond what is justified by events, frenzied and yet ritualized, since everyone cares about not being misinterpreted. This is the way to understand the Red Scare after World War I, the widespread paranoia during the Great Depression, the McCarthy episode during the Cold War, and anti-immigrant chauvinism during every major war.

But the main interest of the cooperation game is in showing the various ways in which a law can have both behavioral and hermeneutic effects. (Recall from Chapter 2 that the hermeneutic effect of a law is its effect on people's beliefs about whether an action reveals a type, like whether informing on Communists is the action of a good type or a bad type.) First, the law can modify the cost of sending a signal. Second, the law can modify the payoffs from cooperation. Third, the law can modify people's beliefs about the proportion of types in the population. Fourth, the law can modify the norm entrepreneur's payoff from constructing a signal or the law can construct a signal itself. These four effects in combination may produce a change in the equilibrium. The behavior in this new equilibrium represents the law's behavioral effect, and the beliefs in this equilibrium represent the law's hermeneutic effect. The following examples illustrate these phenomena.

Honoring and Desecrating the Flag

One way to signal is to show respect for the flag. I will call "saluting the flag" any of a range of actions, including actually saluting the flag when the opportunity arises, displaying the flag, following the rituals that govern care of a flag, and so on. I will call "denigrating the flag" any gesture perceived as disrespect for the flag, from ignoring the flag when one should show respect for it to burning it on the steps of the courthouse. Of course, flags are cheap, and putting one up on a flagpole is cheap, but the costs of observing all of the rituals surrounding flag owning and saluting quickly mount, and anyway this is just one of many signals that people use.

The cooperation game shows why enthusiasm for the flag waxes and wanes. In times of crisis the cost of being ostracized is so great that no one would risk the punishment that might result from deviation from an equilibrium in which everyone respects the flag. In times of security the cost of being ostracized is relatively low, so bad types do not bother to salute the flag. But if people will cooperate with people who do not salute the flag, good types will not incur the cost of saluting the flag. Therefore, no one will salute the flag. In times of tension, separating may occur, as only the good types find it profitable to salute the flag.

Does it make sense for people to believe that those who fail to show respect for the flag are not good cooperative partners? It depends. In the United States a combination of accident and design has created an association in the minds of almost everyone between a particular pattern of stars and stripes and a cluster of beliefs. This cluster of beliefs includes at its core a commitment to the political structure of the United States of America. Because of accident and design, then, the flag is a focal point. If people see someone displaying the American flag on his house, they assume that he intends to express his commitment to the United States. In contrast, if people see someone displaying a flag that bears a design that they do not recognize, they will not understand the meaning that the person is trying to convey.

As noted earlier, because saluting the American flag has a relatively clear meaning, one does not think that a person displays a flag just because of a peculiar aesthetic taste. One believes that the person who displays the flag incurs costs in doing so, and because these costs are not exceeded by any intrinsic benefits, they must be incurred only for reputational purposes. Because the person who displays a flag could only do so for reputational gains if he has a low discount rate, that person must belong to the good type. He would make a good cooperative partner. It is important to emphasize that the reason that he would make a good cooperative partner is not that he necessarily has the patriotic beliefs that we associate with flag waving. The buyer of widgets cares little about whether the seller is a patriot or not, however that term is defined. The buyer of widgets is attracted to the flag waver because the cost of flag waving implies that the flag waver has a low discount rate.

The disjunction between the two meanings of flag waving—that one has patriotic beliefs and that one has a low discount rate—produces some peculiar phenomena. Because burning a flag (and related conduct) is as costly as waving a flag, burning a flag can also serve to reveal that one has a low discount rate. Indeed, because burning a flag is a pithy expression of one's rejection of the values of those who honor the flag, it is particularly expensive, as it reduces one's opportunities to deal with those who respect the flag. Therefore, burning the flag can be an effective signal if one belongs to a group that rejects the values of the majority. It is an effective signal because it is costly. It is, in addition, an effective commitment device because the resulting ostracism from the dominant community increases the cost of being ostracized from the subcommunity, and thus reduces the incentive to free-ride. In a separating equilibrium both the flag waver and the flag burner have low discount rates, and are good cooperators within their respective groups; the people who show no reaction to the flag one way or the other are the bad types in each group. This phenomenon is not limited to flags; it applies as well to the destruction of icons during the Byzantine Empire, the destruction of churches during the Reformation, the French Revolution, the Spanish Civil War, and the destruction of the statues of communist heroes during the revolutions of 1989–1991.

A separating equilibrium may be socially valuable or not, because saluting the flag is an imperfect signal of the propensity to engage in cooperation. Some people will show respect for the flag in order to obtain cooperative gains, but still cheat when the time is right. These people can show respect for the flag, then cheat, because they attach idiosyncratically low valuations to flag waving and other patriotic displays. Other people, who believe that patriotism rightly understood requires non-participation in patriotic displays, will decline to show respect for the flag because of the intrinsic costliness of doing so, and the gains from cooperating with them will be lost. As long as error is low enough that people gain by cooperating only with those who salute the flag, the equilibrium will be sustained. A pooling equilibrium might be superior or inferior to this state of affairs. If everyone salutes the flag because the cost of deviating is extremely high (say, in time of war), while the cost of the signal itself is relatively low, unnecessary and costly signaling will occur but at least the valuable patriotic people who don't like waving the flag will not be ostracized (since even they will wave the flag). But notice that over time people will realize that some flag wavers must be bad types, and that flag waving is not a reliable signal of patriotism. Instead, flag waving becomes a hollow ritual, and this may drive good types over time to abandon this signal—and the norm entrepreneur to create new ones. This is an example of *symbol transformation.*

The state can influence the flag waving game in various ways. First, the state can modify the cost of saluting the flag. Consider a law that not only forbids the burning of flags but requires people to show respect for the flag. Suppose that before the law is passed, a separating equilibrium exists. Good types show respect for flags; bad types desecrate them. The law decreases the citizens' cost of saluting the flag by increasing the cost of the substitutes, desecrating or ignoring the flag. This change might create a pooling equilibrium. Now that it costs so much *not* to send the signal, bad types are forced to mimic the good types and send the signal.

However, the law could have a different effect on behavior. When the cost of *not* sending the signal rises, people may anticipate that everyone—good types and bad types alike—will show respect for the flag. This being the case, people no longer can rely on respect for the flag as an indicator of a person's type. If a person's loss from cooperating with a bad type is high enough, that person will refuse to cooperate with anyone when both types can afford to salute the flag. But if the message senders anticipate this reaction, they will not bother to incur the cost of showing respect for the flag, even though the law has reduced this cost. Why incur this cost, if others are not going to respond by cooperating? To see this more clearly, suppose that rather than punishing people who fail to respect the flag, the law gives monetary rewards to those who do—tax breaks to those who display flags on their houses, show up to parades, and so on—so that these actions are quite cheap (though still more costly than the next best alternative). Here, the model and intuition converge on the prediction that people might respond to this law by refusing to display flags and staying away from parades. The law results in a pooling equilibrium, in which no one shows respect for the flag.

Predicting the law's behavioral effect is all but impossible: it could increase the amount of respect shown for the flag or reduce it.[2] The defender of flag desecration laws, however, might argue that the purpose of the law is to change beliefs, not behavior: to instill in people feelings of respect for the flag. But the hermeneutic effect of the law is likewise impossible to predict. Respect for the flag increases when people increasingly believe that only good types show respect for the flag. If the status quo equilibrium consists of little or no flag waving, and if the law creates a separating equilibrium, then the law might cause people to abandon their belief that flag saluters are odd or idiosyncratic people and accept the belief that they are good types. But if the status quo equilibrium is a separating equilibrium, and the purpose of the law is to enhance people's respect for the flag, the law may well fail. If it produces a pooling equilibrium in which no one sends the signal, people may no longer associate saluting the flag with any character type. If the law produces a pooling equilibrium in which everyone sends the signal, people will believe that everyone who waves a flag may be a good type *or* a bad type. We term this phenomenon *reification.* The law in this case ambiguates the meaning of the symbol, rather than increasing respect for the symbol. If the law's purpose is to enhance respect for the flag, then the law is self-defeating.

The complexity of predicting the effect of a flag-burning law on behavior and beliefs should be evident.[3] One should doubt, then, the popular view that a law against flag burning would have a predictable effect that would be socially desirable. Then why has there been so much support in the United States for a law against flag burning? This question is answered later in this chapter.

Censorship: Governmental and Social

Self-censorship, like respect for the flag, may emerge as a signal of cooperativeness. In the struggle to find ways of distinguishing themselves as good types, people accuse critics of the government of being bad types, while drawing attention to their own support of the government, by implication a signal of their patriotism. The average person may like to criticize the government or feel an obligation to do so; thus, not doing so will usually be a cost. But good types recover their cost through cooperation with others, while bad types do not. The association between self-censorship and patriotism arises frequently, in part because of traditions of self-censorship and deference to authority, and in part because of the threat posed by internecine conflict to national survival in times of war.

The difference between self-censorship and respect for the flag is just a difference in the cost structure of the action. Some people might find self-censorship more costly; others might find respect for the flag more costly. In any event, whether one action, the other, or both actions emerge as signals of cooperativeness depends on the various costs faced by the different types and the extent to which each signal is made focal by tradition and other circumstances.

The problem with self-censorship equilibriums is that they may reduce the well-being of the population. This point is explored at length by Kuran (1995), so I will be brief. When the reputational costs are high enough, serious-minded good types, who care deeply about the state but disagree with prevailing views, may avoid criticizing the government, in which case valuable information is lost.[4] During the McCarthy era, when the cost of criticizing the government (or, at any rate, McCarthy and the policies he supported) became extremely high, many people—good types and bad types alike—stopped criticizing the government, and in this equilibrium not only was socially valuable patriotic criticism lost, so was the valuable exposure of bad types. Although this analysis may seem to presuppose an overly instrumental approach to politics on the part of citizens and politicians, a little reflection ought to persuade the reader that it is accurate. President Clinton may have deeply believed in the equal treatment of homosexuals, but he rapidly learned that efforts to help them would cost him the political support that he needed to accomplish other elements of his agenda. It is common for people to suppress their views in order to refrain from alienating a community that they would like to change in some other respect.

One of the most striking aspects of McCarthyism was that this campaign resulted from McCarthy's entrepreneurial modification of focal points, not from changes in the law. But we can also analyze the effect of a law, such as a censorship law. Imagine that a pooling equilibrium exists: no one censors himself. The state enacts a censorship law. If the law provides for the correct level of sanctions, it creates a separating equilibrium. The law has two effects one might care about: it reveals bad types (the *internal* effect) and it discourages criticism of the government (the *external* effect). This is characteristic of all the laws under consideration. The flag desecration law reveals bad types and it encourages respect for the flag. The censorship law may be more effective than the flag desecration law at exposing bad types, but it may also cause more harm than the flag desecration law, because speech has valuable external effects that responses to the flag, for the most part, lack.

The state can also produce or maintain separating equilibria or pooling equilibria by influencing the beliefs about the fraction of bad types in the population. Suppose that people believe that almost all people are good types and cooperate with everyone, so no one sends the signal of not criticizing the government. The government now warns that a lot of people are in fact bad types. If the government is persuasive, people may refuse to accept new cooperative partners (or, more realistically, they may take precautionary measures that lower the gains to each other); in response, the good types distinguish themselves as good types by declining to criticize the government. If bad types would not be able to recover their costs, they will not mimic the good types; a separating equilibrium is produced. Because people learn the types of only the senders of signals, they will have no grounds for later revising their beliefs despite their inaccuracy.

Propaganda can be analyzed in another way. In the norm entrepreneur (NE) game the state can take the role of norm entrepreneur by issuing propaganda. By issuing propaganda the state creates an opportunity for people to signal their patriotism or subversion. The signal is self-censorship—resisting one's impulse to disagree with the propaganda. The more baldly untruthful the propaganda, the more clearly does a person signal his patriotism by declining to disagree with it. Official lies are in this way like a flag or a national holiday: they provide the opportunity for demonstrating one's loyalty to the government.

We can approach the problem of government censorship and self-censorship from another angle. Distinguish two kinds of speech: speech that the average person finds offensive, and speech that criticizes the government. Currently, the average American seems to be offended by pornography, mockery of religious beliefs, criticism of the American system of government, and denigrating comments about women and minorities. In other countries, ordinary people are often offended by pornography, mockery of religious beliefs held by most people, criticism of the basic political system and of the dominant ethnic group, and so on. Call this kind of speech (rather awkwardly) "socially offensive speech." Call speech that criticizes the existing government, as opposed to the political or constitutional system, "politically offensive speech."

Any state might censor either, neither, or both kinds of speech. Let us say that governments that censor socially offensive speech engage in "s-censorship," and that governments that censor politically offensive speech engage in "p-censorship." Totalitarian states like the Soviet Union engaged in s-censorship and p-censorship. Many mainstream democracies engage in only s-censorship, and some (like the Netherlands) do not engage in any censorship. A few states engage in p-censorship but not s-censorship: apparently, Serbia is such a state (Hedges 1998). These distinctions allow us to propose a few hypotheses.

1. An obvious hypothesis is that s-censorship can exist in a well-functioning democracy, but p-censorship cannot. Because the average person is offended by socially offensive speech, an ordinary democracy will, barring constitutional constraints, implement that view through law. But except when leaders are exceptionally popular, ordinary people are not offended by criticisms of the existing government.
2. P- and s-censorship converge at times of great national peril, such as war. During war, the median voter wants both to see his ordinary preferences about social speech respected, and to preserve the government from criticism (at least as long as he thinks that the government is doing a good job).
3. Individuals who violate p-censorship laws and are punished receive more status and fame, and become more influential, than citizens who violate s-censorship laws. By "influential," I mean able to cause changes that affect a lot of people, like the collapse of the government or changes in social mores. Compare Sakharov and Larry Flynt. No doubt many pornographers went to jail in the Soviet Union and we have never heard of them.
4. Individuals with extreme preferences (or strong commitments) are more likely to be influential under p-censorship than under a regime of

free speech. This is the martyr idea: Sakharov violates the law and becomes influential in the Soviet Union; but people in the United States cannot violate p-censorship so there is no one of equivalent stature (compare Hirschman 1982, pp. 105–06).

5. When self-censorship (for example, political correctness or religious orthodoxy) exists, people will ask the government to engage in s-censorship that supports it (for example, hate crime codes, religious orthodoxy laws); but s-censorship can inadvertently destroy the equilibrium in which self-censorship prevails by destroying the signaling value of the original form of self-censorship.
6. Governments will attempt to justify p-censorship by blurring the line between p-censorship and s-censorship. For example, governments will claim that because they protect public morals (s-censorship), public attacks on the government will lead to the destruction of public morals, so p-censorship is also justified. This may explain why it is a bad idea to distinguish political speech and non-political speech for purposes of first amendment analysis.
7. P-censorship and s-censorship create different kinds of social phenomena. People may violate (or circumvent) p-censorship in order to show that they oppose the government. People violate s-censorship to show that they oppose social mores. Both may be a way that people match up with each other to form social groups. It is the difference between the radical students of the 60s (political) and the hippies (cultural).

Voting and Other Forms of Civic Participation

Rational choice theorists have not produced a satisfactory explanation of why people vote. Given the vanishingly small chance of influencing the outcome of an election and the relatively high and certain cost of taking time from work and standing in line at the voting booth, one would expect people never to vote. Yet people do; this is the so-called voting paradox. Positing a "taste" for voting (or, in some approaches, for "self-expression") and other efforts to play with utility functions have not produced satisfactory results (Green and Shapiro 1994, p. 70). One might assume a social norm in favor of voting, and work from there, but this strategy transforms the question, Why do people vote? into the question, Why does a social norm require people to vote?

Because an individual's vote does not influence the outcome of an election, it should be considered a purely symbolic act, like displaying the American flag. Thus, voting can be analyzed as signaling behavior. The signal here is the act of voting in the voting booth, not the vote in favor of one person or another. Voting is observable, even if the casting of the vote is not: one's friends, associates, and family members know that one voted, because one took time off work or simply told them that one did when duplicity would risk one's reputation. The act of voting is costly both for the good type and for the bad type, but the good type recovers his costs through repeated rounds of cooperation with another person, while the bad type does not. In the separating equilibrium the voter is believed to be a good type, and the non-voter is believed to be a bad type. This explanation turns the voting paradox on its head: voting functions as a signal precisely because the costs exceed the material gains. If voting were profitable, as would be the case if people were paid to vote or severely punished for not voting, or if everyone derived "expressive utility" from voting, then good types and bad types would all vote. The motive for voting, on this theory, is not to satisfy some taste for voting, or for expressing one's views, or even for helping one's country, but to obtain cooperative returns from other private actors. Patriotism is relevant only as a theory of the origin of the psychological association of voting and cooperation.

In a separating equilibrium the more cooperative people vote and the less cooperative people decline to vote. In the United States many people disapprove of those who do not vote and who admit it. This reaction is inexplicable under the theory that posits a taste for voting, whereas it follows from the signaling model: the disapproval expresses people's view that you may not be a trustworthy person. Times of national emergency could stimulate a pooling equilibrium in which everyone (or nearly everyone) voted. The main reason for voting at such times is not that one's single vote now matters (it matters less than ever); but that it now matters if people think one is a bad type.

Let me mention a few pieces of evidence that are consistent with the signaling theory of voting. First, the fact that voting increases with wealth and education (Leighley and Nagler 1992, Durden and Gaynor 1987) supports the signaling theory but contradicts the taste theory. An action can serve as a signal only if it is costly, and the cost of voting increases with one's opportunity costs. In contrast, under the taste theory people would substitute to actions that satisfied other tastes as the cost of satisfying the taste for voting increased.

Second, voting increases with membership activity in organizations like parent-teacher associations, charitable organizations, neighborhood organizations, business organizations, interest groups, and unions[5]—all organizations that require cooperative behavior from their members. The taste theory does not explain why people who belong to organizations would vote more often. The signaling theory suggests that the same sort of people—good types who seek cooperative partners—would send signals by voting and by incurring the cost of entering an organization.

Third, respondents' frequent exaggeration of their voting behavior to pollsters suggests an embarrassment about admitting failure to vote. People who have a taste not to vote should be no more embarrassed than people who have a taste not to eat apples. But embarrassment results because the respondent knows that an admission that he does not vote injures his reputation (Presser and Traugott 1992; see also Hasen 1996, pp. 2160–61). He is failing to send the appropriate signal. To be sure, it is unlikely that he cares whether an anonymous caller believes that he belongs to the good or bad type; but he also has no reason to disclose to the caller the truth about his type.

Fourth, the fact that people disapprove of those who do not vote removes this behavior from the realm of taste. If people do not disapprove of those who dislike the taste of apples or oranges, why should they disapprove of people who lack the taste for voting? Under the signaling theory, disapproval is an expression of people's inclination to avoid those whom they do not trust.

Fifth, voter ignorance is more consistent with the signaling theory than with the taste theory. One study found that votes were cast by 43 percent of a population of people who saw no difference between the candidates (Mueller 1989, p. 357, citing Brody and Page 1973). The taste theory would predict that these people would not vote, since they could not gain from one candidate winning rather than the other. But under the signaling theory it does not matter who wins; what matters is the fact of voting. Indeed, the signaling theory but not the taste theory can explain the strong showing in a primary election for the Democratic nomination for a seat in the United States Senate by a candidate who happened to be dead ("Dead Woman Forces Runoff in Senate Race" 1998). Other studies show ambiguous evidence for the proposition—crucial to the taste theory but indifferent for the signaling theory—that voting increases as the probability of affecting the outcome increases (Mueller 1989, p. 358). To be sure, voters will vote for the candidates whom they prefer when they are in the voting booth—why wouldn't they?—and their votes will often be correlated with their interests because of general information they have.[6] This fact is consistent with both models.

Finally, people who do not vote tolerate unconventional lifestyles more than people who vote (Crotty 1991, p. 10). The signaling theory, but not the taste theory, explains this association. It is the consequence of the incentive of the good type to distinguish himself by conforming to social norms.

It is always difficult to generate predictions from signaling theories. The previous examples show that the facts about voting are consistent with the signaling theory and not with the taste theory, but the facts are also consistent with an indefinitely large number of other possible theories of voting behavior. If one streamlines the signaling theory by committing to an equilibrium refinement, one can generate some hypotheses that are worth testing, including (1) the rate of voting will increase during times of tension; (2) the rate of voting will increase when the act of voting is made increasingly observable; (3) the rate of voting will be unrelated to how much information people have about the candidates (as long as one's choice remains anonymous). The first hypothesis is plausible, but I know of no study that tests it. Anecdotal evidence for the second hypothesis is the high voter turnout in Italy where the names of non-voters are posted in a public area (Hasen 1996, pp. 2169–71). Anecdotal evidence for the third hypothesis is the well-known fact that people sometimes vote for candidates on the basis of the sound of their names, knowing nothing about them.

The hypotheses just mentioned are not valid when the cost of voting becomes too high. If times are tense enough—say, because of riots in the streets—no one will bother to vote. So another way to distinguish the signaling theory and the taste theory is to measure voter turnout as the cost of voting increases over a sufficiently wide range. Under the taste theory, as the cost of voting increases, people should vote less. Under the signaling theory, one would expect the rate of voting to increase initially, peak, and then decline. The rate of voting begins at a low level because the cost of voting is not high enough to serve as a signal; it increases, perhaps sharply or discontinuously, when the cost of voting is high enough to serve as a signal; then when the cost rises above a certain threshold, people will stop voting because they cannot afford to. Again, because the reputational gains from voting will disappear as soon as a threshold number of people stop voting, the decline should be precipitous. Cost could be direct, like a poll tax, or indirect, in the form of opportunity costs that rise when queues lengthen and voting becomes more inconvenient for other reasons.

Hand-wringing about low voter turnout is often scoffed at, because it could mean that people are relatively satisfied with the political status quo. The problem, however, is that because of peace and prosperity, it matters less if one is believed to be a bad type, so people do not bother to vote as a signal of patriotism. If, as a result, voter turnout falls, the variance of outcomes will increase, resulting in the election of politicians whose agendas do not reflect the interests of citizens. However, equilibriums in which many or all people vote are not necessarily desirable, either. People who vote solely for reputational reasons will not take their vote seriously, failing to inform themselves about the various candidates and voting instead on the basis of the mellifluousness or familiarity of the candidates' names. This means that legal efforts to increase the rate of voting, including Italy's law mentioned above and laws in other countries that fine people who fail to vote, do not necessarily result in any public good: more voting, but not more informed voting.

Some Positive Implications

In predicting the effect of a law, one must know the way it modifies the cost of the action that serves as a signal, the payoffs from cooperation, and people's beliefs about the type of person who sends the signal. One must also take account of further complications. First, the effect of a law depends on the status quo equilibrium. A censorship law imposed on a pooling equilibrium in which everyone sends the signal of self-censorship may have no effect on behavior; but imposed on other equilibriums, it may cause a significant change in behavior. Second, the effect of a law even on a given equilibrium may be unpredictable: as we saw earlier, a law that reduces the cost of a signal may produce a pooling equilibrium in which everyone sends the signal, but it also may produce a pooling equilibrium in which no one does.

Two other complications are discontinuity and symbol transformation. To understand the problem of discontinuity, imagine a pooling equilibrium in which everyone discriminates against communists. Assume that "tastes" for discrimination are distributed uniformly: many people would prefer not to discriminate but suppress their inclinations to avoid being ostracized. A law is enacted that punishes those who discriminate. If the sanction is small, it is unlikely to affect behavior: the cost of ostracism exceeds the cost of the sanction. As the sanction is gradually increased, discrimination does not change by much. As long as the reputational sanction exceeds the legal sanction, behavior is unaffected except for those with the most extreme preferences. But at some threshold, when the legal sanction exceeds the reputational sanction, the amount of discrimination will decline discontinuously. The reason is that once people with the stronger preferences against discrimination deviate from the pooling equilibrium in sufficient numbers, the reputational sanction disappears, and all the people who would cooperate with the members of the minority group but for the existence of the reputational sanction will stop discriminating (compare Bernheim 1994, Kuran 1995). This phenomenon means that a law that has a small sanction (or even no sanction, like an announcement by an official) may have a disproportionate influence on behavior, and a law that has a large sanction may have little or no influence on behavior if by increasing the cost it enforces or nullifies the signal. Social norms are brittle.

Symbol transformations occur because exogenous changes cause old signals to fail, eliminating people's ability to distinguish good types and bad types and thus giving good types and norm entrepreneurs an incentive to discover new signals to replace the old signals. One example, which is presented in stylized form, comes from the McCarthy period. At roughly the same time that McCarthy was making the connection between communism and subversion, others were asserting a connection between homosexuality and subversion.[7] Because of the Kinsey report and the social and demographic dislocations caused by World War II, people rapidly became aware that a lot of people engaged in homosexual behavior, just as a lot of people supported communism. Because refusing to cooperate with identifiable homosexuals is costly, discrimination could serve as a signal of cooperativeness. But for this signal to function properly, it was necessary for people to believe that discrimination expressed a desire to signal cooperativeness rather than a moral conviction, prejudice, or taste. Norm entrepreneurs had to draw the connection between homosexuality and subversion, and this was done in a variety of ways—by appealing to traditional moral and religious antipathy to homosexual behavior while claiming the importance of unity for national security, and by claiming that homosexuals corrupted youth on whose "manliness" the nation relied. Sweeping legal and nonlegal attacks on homosexuality paralleled the mostly unofficial attacks on communists (D'Emilio 1983).

Symbol transformation occurs when one signal (self-censorship) fails to expose bad types, resulting in the switch to or the addition of another signal (discrimination against homosexuals). The irony is that if any American citizens posed a threat to the United States in the 1950s, it is more likely that communists and other political critics did than homosexuals, and yet discrimination against homosexuals proved to create a far more powerful equilibrium than discrimination against communists. The reason is probably that this country's traditions prior to the 1960s supported political freedom more than sexual freedom, so self-censorship was a less reliable indication of patriotism—indeed, could be interpreted as a failure of patriotism—than discrimination against homosexuals.

Normative Implications: The Behavioral Effect

The examples illustrate some generalizations. First, there is a tension between the motive that causes a person to take an action and that action's contribution to a public good; and there is a tension between two kinds of public goods produced by an action—the revelation of information, on the one hand, and the production of some other public good, on the other. For example, a person may vote or engage in self-censorship from the motive of signaling his low discount rate. Voting, happily, produces an external public good (revelation of political preferences); self-censorship produces an external public bad (concealment of political preferences). Both kinds of action produce an internal public good, namely, the exposure of bad types. But even here the cost of signaling may exceed the benefits of that information. There is nothing intrinsically socially beneficial about signaling: sometimes it is beneficial, sometimes it is not.

Second, signals are crude, and even equilibriums that might be considered broadly socially beneficial result in some undesirable behavior. Some politically uninformed people vote to avoid reputational sanctions. It would be more desirable if they followed their non-reputational preferences and did not vote. When self-censorship equilibriums arise, political stability is established, and this may be important in times of war, but good types with valid criticisms of the government will be silenced along with the bad types.

Third, when signaling equilibriums are beneficial, we can expect them to occur too rarely. The reason is that here the signal is a public good. (This is not true in standard signaling models, but they fail to account for the ambiguity of actions that may serve as signals.) To be established, everyone has to recognize that certain actions are cheaper for some types than for others; but because the benefits of the signal are enjoyed by everyone, whereas the costs of making this connection are born by a few (the norm entrepreneurs, the gossips), there will be too few kinds of signals that are available for use (compare Lessig 1995). For similar reasons, harmful equilibriums occur more rarely than they would if signals were not produced through collective efforts.

The Hermeneutic Effect: The Creation of Social Meaning and the Problem of Reification

When the law changes a separating equilibrium into a pooling equilibrium in which everyone sends the signal, people stop associating the signal with cooperativeness. If we care about the "social meaning" of the action that serves as a signal, this result may be undesirable. This is the difference between raising an American flag in a hostile foreign country and putting it on your house in an American suburb, or the difference between criticism of the authorities in a dictatorship and criticism of the authorities in a democracy, or the difference between wearing a pink triangle on a college campus in the 1970s or early 1980s and doing so in the 1990s. Because everyone or almost everyone issues the signal, it no longer distinguishes some people from others. The "social meaning" of an action (Lessig 1995), which can be defined as the belief that the average person has about the type of person who engages in that action, has become *reified.*

When the law changes a separating equilibrium into a pooling equilibrium in which no one sends the signal, the signal disappears. Veterans of wars complain that people do not take the flag as seriously now as they used to. Then, the signal, though reified, was not meaningless: few people failed to salute the flag, but those who did fail to salute the flag were necessarily bad types, so those who saluted were at least possibly good types. Now, in many circles saluting the flag is almost meaningless. The observer assumes that the person who salutes the flag is old-fashioned or even a bit dotty—a person with strange tastes. When so few people issue a signal that the observer no longer associates that signal with a particular type, the social meaning has been *destroyed.* But as the Texas Court of Criminal Appeals noted in its opinion striking down Texas' flag desecration law, this has not yet happened to the flag, and would not happen until it had lost its ability to "rouse feelings of unity and patriotism" and instead had become "a meaningless piece of cloth" (Goldstein 1996, p. 63).

When the state converts a pooling equilibrium in which no one sends the signal into a separating equilibrium, it produces social meaning. An action that previously had little significance now has a great deal. Saluting a piece of cloth displaying stars and stripes had no significance prior to the Revolutionary War; later, it would have a great deal of meaning. Discrimination against those who engage in homosexual behavior was seen initially simply as a reaction to a practice believed to be immoral; after state endorsement it is seen as an expression of patriotism. A social meaning is *created.*

But for all the emphasis on the role of the state, social meanings can emerge and disappear spontaneously, and often in the face of state efforts to regulate them. To see why, suppose at time 1 there is a separating equilibrium, and good types salute the flag just to show that they are good types. At time 2 the state enacts a law that punishes people who fail to salute the flag. At time 3 there is a pooling equilibrium, created by the bad types' desire to avoid the punishment. But at time 3 saluting the flag is no longer a reliable signal of patriotism. If many people salute the flag just to avoid the sanction, then those who salute the flag are not necessarily good types; the salute becomes reified. In the patriotism game, both types continue to salute the flag, the good types fearing that if they fail to salute they will be mistaken for bad types. But over time many people, good types or not, will begin to recognize saluting the flag as the empty ritual that it has become. Saluting the flag becomes an embarrassment, because everyone knows that people salute the flag just to avoid legal punishment. Some will conceal their embarrassment behind a mask of irony, but under such conditions the meaning of the salute may eventually *flip,* becoming instead a signal of fear of legal punishment rather than a signal of patriotism (compare Kuran 1995). The person who salutes is slavishly obedient, fearful to offend the authorities or other people; the person who declines to salute has integrity and independence. Failing to salute the flag becomes a better signal of loyalty than saluting the flag (and I believe that in some circles this is the case).

The *politicization* of behavior occurs with the creation of a law that requires people to engage in some behavior in which they had previously engaged voluntarily. People already salute the flag or pray at ceremonies; then a law is created that requires exactly the same behavior. At first sight, one would expect the law not to affect behavior, perhaps even to intensify it. But the law may flip the signal, so that the sender fears that others will think that he engages in the behavior to comply with the law, rather than to express genuine patriotism or religious fervor. The result may or may not be that people stop engaging in the behavior—that depends on the size of the sanctions and other factors. The important conclusion for present purposes is that politicization destroys important social meanings by legally compelling behavior that derives its meaning in part from the fact that it is not required by law. This argument is an analogy to the argument that the *commodification* of goods and services through the market destroys social meanings when a behavior had derived its meaning from the fact that it is given freely (compare Radin 1987). Commodification and politicization are mirror images.

Should the state "regulate social meanings," as advocated by Lessig (1995)? The problem is that the results of government efforts to change or sustain symbols, whether through legal devices or official exhortation, are unpredictable. Government efforts to change signals can lead to a strengthening of symbols that the government sought to change (a possible example is President Clinton's unsuccessful efforts to increase acceptance of gays in the military), or to reification of the desired symbol (a possible effect of a law banning flag desecration). And when government efforts, whether deliberately or not, destroy or reify existing symbols, norm entrepreneurs will propose new symbols that may have even worse effects than the old ones. When searching for examples of successful government-led efforts to exploit symbolic behavior, one repeatedly finds the most vivid triumphs in the histories of fascist and totalitarian states—not attractive models for the United States, a country in which the most striking successes in norm entrepreneurship have been achieved by non-governmental movements such as civil rights, feminism, and religious evangelism.

These observations are all at variance with the current enthusiasm for the idea of deliberative democracy. Deliberation is a good thing, if it means one thinks before one acts. But stronger claims are hazardous. Because of equilibriums in which people who express dissenting views are considered bad types, dissenters will rationally conceal their views, and ordinary people will be deprived of information that would enable them to correct false beliefs. Political "deliberation"—in the sense of the actual public debate that exists in the real world, not the hypothetical, stylized debates that occur in philosophical thought experiments—is as likely to harden such an equilibrium as to undermine it. Then people act on these false views by voting for politicians who will implement them as law.[8] Examples are too numerous to list; a few include the Love Canal scare and other regulatory bubbles (Kuran and Sunstein 1999), the red scares in the United States, and war enthusiasms all over.[9]

Endogenizing the State

There is another problem with government norm entrepreneurship. To see this problem, suppose that everyone engages in self-censorship in order to avoid being labeled a bad type. This pooling equilibrium does not serve the interest of most people, but every citizen is afraid to deviate. Now we might say that here is a collective action problem that the government could solve: for example, by subsidizing the publication of newspapers and journals, giving politicians free air time, enacting special legal immunities against charges of libel and slander, and granting other privileges and subsidies. The question is why we should expect the government to engage in any of these actions. The problem is not just that government officials may enjoy the lack of criticism. The problem is that government officials will not propose these laws, lest their support for them be taken as a signal that they are bad types. Citizens will not lobby the government to enact these laws, because they fear that their lobbying—a violation of the self-censorship norm—will be taken as a signal that they are bad types. We cannot expect the government to change socially undesirable social meanings when these meanings are sufficiently powerful.

These considerations return us to the question raised at the start of this chapter: why do people seek laws against flag burning when flag burning causes no "real" harm? The answer is that citizens signal their patriotism not only by engaging in patriotic activity but by lobbying or at least passively showing support for laws that punish people who engage in unpatriotic activity. Since the signal, political support, is still relatively cheap, a pooling equilibrium in which everyone sends the signal can result. Once *support* for a law against flag burning is taken as a signal of one's patriotism, elected government officials cannot afford the political consequences of opposing such a law. Government officials do not stand outside the signaling game. They, like citizens, are prisoners of symbols when the symbols are sufficiently powerful. Truman and Eisenhower were powerless to resist McCarthy at his height, because any effort to criticize McCarthy would have been interpreted as a sign of the presidents' pusillanimity, even sympathy, toward America's enemies.

A further problem with government-led norm entrepreneurship is that it produces wasteful competition for government resources. Competing groups treat the government as an instrument for conveying their symbols. An important example is the placement of religious symbols on public property. The problem is that observers understand that the symbol represents the successful lobbying efforts of one religious group, and thus that this religious group has significant political influence. This inference causes people to assume that members of the religious group exist in great numbers or that they have great power—in either case, they are likely to be attractive (or unavoidable) cooperative partners. Thus, attempts by members of the majority to use discrimination against members of this group as a signal of loyalty or patriotism will fail. It is thus in the interest of religious groups to compete for influence over the government's decision to use religious symbols. Constitutional restrictions on establishment deter wasteful competition among religious groups over the use of the government as an instrument for recruiting members through its endorsement of their beliefs.[10] The free exercise clause and equal protection rules protect members of minority religions against the kind of discrimination that would stimulate the pursuit of political power in the first place.

8

Racial Discrimination and Nationalism

The analysis in the prior chapter can be applied to the question of how the law influences discrimination on the basis of race, ethnicity, ancestry, and similar categories. To answer this question, one needs a theory of why people discriminate.

Why Do People Discriminate?

There is no satisfactory economic theory of why people engage in racial discrimination. Becker (1971), Epstein (1992), and others assume that people have a "taste" for discrimination. It is not clear whether this assumption is a methodological convenience or is intended to reflect a psychological theory of behavior. The better reading of the literature is that the assumption is a methodological convenience that allows one to predict the effect on discriminatory behavior that would be produced by changes in laws and other exogenous factors while one remains agnostic about the source of people's racial preferences. Some writings, however, imply that people's racial preferences are natural or fixed, and for expository convenience I will refer to this view as the "preference theory."

The preference theory is implausible, and is inferior to a signaling theory of why people discriminate. According to the latter theory, discrimination against people with salient and immutable characteristics that systematically differ from those of desired cooperative partners serves as a signal to the latter that one has a low discount rate. This theory, unlike the preference theory, treats the preference to discriminate as endogenous: people derive utility from discriminating against members of other races only to the extent that such discrimination earns them reputational gains in cooperative relationships.[1] The theory can also be contrasted with theories of statistical discrimination, which hold that people discriminate when race is a proxy for an undesirable characteristic (Arrow 1973). These theories, which are usually proposed to explain workplace discrimination, assume that the relevant hidden information is a worker's quality, and imply that when people engage in statistical discrimination, workers respond by underinvesting in human capital (Arrow 1973, Coate and Loury 1993, Cooter 1994a, Schwab 1986). Akerlof (1984, ch. 5) treats as exogenous exclusionary social customs and shows that in small markets such customs can be stable against efforts by nondiscriminating entrepreneurs to enter.[2] McAdams (1995) argues that racial discrimination results from status competition.

The preference theory predicts that discrimination on the basis of race will occur as long as people have a taste to discriminate, and the cost of satisfying that taste is not too high. The theory makes several interesting predictions about the employment and compensation of minorities, on which see Becker (1971).

The predictions of the signaling theory are different. First, discrimination is more likely to occur as tensions rise. People seek to signal to each other that they belong to the good type, because cooperative partners are in high demand. Only when there is little scarcity and near-perfect information will discrimination serve no purpose at all.

Second, the people who are the targets of discrimination share immutable characteristics that are observable (like skin color) or discoverable (like religion or ancestry), and that differ from the characteristics of the desired partners. The characteristics must be immutable; otherwise, people who are targets of discrimination would simply change them.[3] Notice that a person's race is a function of those characteristics that enable discrimination. It is not that people discriminate against one because one belongs to a different race; it is that one belongs to a different race because people discriminate against one on the basis of immutable characteristics that one shares with others.

Third, discrimination against a "race" must have the right cost structure: it cannot be too cheap or too expensive. It will be too cheap if the people who share the immutable characteristics are so few or poor or unskilled that refusing to deal with them is not costly. It will be too expensive if the people who share the immutable characteristics are so numerous or wealthy or skilled that one cannot prosper unless one deals with them. Like voting, racial discrimination will occur only when the cost of discrimination is neither too high nor too low. And it should be the case that there is not a superior form of signaling (like gift-giving, patriotic displays, or religious discrimination) that produces the same results at less cost.

Fourth, discrimination against a "race" must be recognized as a signal of cooperativeness rather than believed to be the satisfaction of some idiosyncratic taste. If people believe that those who engage in discrimination are motivated by idiosyncratic tastes, then they will not believe that people who discriminate are necessarily good types. Then there would be no norm of race discrimination, any more than there are norms about eating apples or oranges. The modern forms of racial discrimination occur only when people believe that those who engage in discrimination are all or mostly people with low discount rates. The question is, how could such a belief get started?

As the discussion of focal points in Chapter 2 suggested, historical accident is an important source of the beliefs that sustain discriminatory equilibriums. Sometimes, a group becomes a focal point of discrimination because members of the group have social or ancestral connections with a group that is unambiguously a threat to the majority group. Examples include Americans of Japanese origin after the bombing of Pearl Harbor, and people of Serbian ancestry living in Croatia or Bosnia. The theory does not require that members of the out-group *really be* a threat to members of the in-group. It requires only that the association be plausible to many people, given their various cultural assumptions and contemporaneous events.

Sometimes, a minority group becomes the focal point of discrimination because its members compete with members of the majority group for scarce resources. If black migration threatens the power of whites in a local labor market, then whites can demonstrate their loyalty to each other by discriminating against blacks. The person who engages in discrimination can plausibly claim that he does so in order to aid members of his group, not in order to satisfy a preference. Alternatively, if a minority group prospers while the majority does poorly, and the source of the minority group's prosperity is not understood but has some connection to the majority's problems, discrimination against the minority may become focal. Jewish success in the credit market and other important markets in European history may have caused people to believe that Jews had political power out of proportion to their numbers. This, along with the long history of religious antagonism, made Jews a focal point for discrimination among Christian Europeans. These phenomena are self-reinforcing: when members of an out-group prosper, they become suspect, and members of the larger society discriminate against them to show their loyalty to each other; but this makes it even more important for out-group members to rely on each other, enhancing their trust and their mutual gains, leading to further discrimination against them by the in-group.

It seems not to be the case that *any* immutable physical characteristic, such as height or hair color or eye color, can serve as the basis for discrimination. A possible reason for this is that the characteristic cannot be too continuously distributed among different people, for then it would be impossible to draw a line between those who have it and those who do not. If the line between the in-group and the out-group is fuzzy, then discrimination against a member of the out-group is difficult to engage in and difficult to perceive, so it cannot serve as a reliable signal of loyalty to the in-group. Discrimination targets characteristics that are systematically different between groups of people, such as skin color and ancestry, which occur as a result of the migration of a group whose members have intermarried over a long period of time.

If discrimination against members of racial groups serves as a signal, then the usual array of equilibriums become possible. In a pooling equilibrium everyone or almost everyone discriminates against members of the out-group. In the old South whites who felt sympathetic to blacks feared associating with blacks in a respectful manner, because other whites might infer that they did not belong to the cooperative type. In a separating equilibrium many people discriminate against members of the out-group, but many people do not. The latter either have lower discount rates or a stronger interest in associating with the members of the out-group.

Discrimination against groups out of purely self-interested motives is often rationalized. A common rationalization that supports discrimination is racial or ethnic superiority, and such rationalizations are endlessly supplied by norm entrepreneurs, of which the Nazis are the preeminent example. The theories are mystical, based on discredited science, supported by no evidence (especially the insistence on the importance of the purity of blood), inconsistent with each other, or just plain weird (Mosse 1978). And yet every ethnicity has its norm entrepreneurs, and has its stories, and has its believers. The reason is that just as one shows one's patriotism in a totalitarian country by endorsing its obviously wrong propaganda (the more obviously wrong, the more effective the signal), one shows one's patriotism in a fascist country or one's loyalty to an ethnic group by participating in collective discrimination against outsiders even when one likes them and even when they are defined in an obviously, even embarrassingly, arbitrary way. What counts as a signal depends on the importance of cooperation for individuals (which rises and falls with fluctuations in economic and political circumstances), the historical accidents that create associations between certain kinds of behavior and certain kinds of people, and the relative costs of the actions that are made salient by these associations. The signaling theory shows how ethnicity and race are inventions that respond to the demand for criteria that facilitate cooperation against potential threats.

The signaling theory is superior to the preference theory for many reasons. Let me discuss a few briefly.

1. The signaling theory follows from the standard assumptions of econom-

ics. It assumes that people's behavior is consistent with their preferences, while taking no position on what those preferences are. The preference theory does not explain why people would not want to associate with people who have systematically different physical characteristics. Indeed, it makes a mess of the whole idea of difference. It is a natural consequence of theories of evolutionary biology that people would have a preference for food, or for clothes, or for interaction with people who are healthy or who have symmetrical features, but not that people would want to avoid people who have some different physical characteristics. Much discrimination occurs even though the targets of discrimination are physically identical, as a class, to the discriminators, and differ only in immediate ancestry or in their cultural practices. Medieval Spaniards discriminated against Jews, even though Christian Spaniards and Jews had intermarried for generations. Serbs, Croats, and Bosnians have divided on ethnic grounds despite the absence of a genetic basis for these divisions. And savage conflicts on the basis of religious and class differences have split members of ethnic groups for hundreds of years. What counts as an ethnic distinction for the purpose of conflict, indeed whether an "ethnic" trait or some other trait will be focal, emerges from deeper forces. Whether the British and Germans should count as one ethnic group (Aryans) or two or more (Scottish, Welsh, English, Saxon, Bavarian, and so on), or whether people in these countries divide along different lines (religion or class), depends on what forms of shunning are—and have been historically—*convenient.* Nineteenth century philosophers of race frequently assumed that the "Aryans" were the preeminent race, but argued about who counted as Aryans, and proposed groups as divergent as east Indians, Iranians, and, of course, northern Europeans (Mosse 1978). In pre–World War II Germany only the last category was politically convenient.

Even if it is true that people do have associational tastes based on systematic physical differences, one should assume that, like any other tastes, they are normally and continuously distributed. Some people like apples very much, some people hate them, and lots of people have intermediate tastes. Similarly, some people should like people with certain characteristics, some people should hate them, and lots of people should have intermediate tastes. As the following observations make clear, if this is so, it cannot be the whole story of why people discriminate.

2. Discrimination against out-groups is usually discontinuous: either everyone or most people discriminate, or no one or few people do. Racial discrimination is usually, in modern times, a behavioral regularity that has reputational implications. We could live in a society in which for most people one's race is not a morally relevant factor in the selection of mates, but we do

not. That we could live in such a society is obvious from the fact that what used to be impregnable barriers (Irish versus Italian, Jew versus gentile) no longer are.

To put this point more clearly, imagine a society consisting of ten Xs and ten Ys, where X and Y refer to different ethnicities, and each group is half male and half female. If people choose mates on the basis of factors other than ethnicity, one would predict that there will be six mixed marriages, two marriages among Xs, and two marriages among Ys. As numbers increase, the curve would remain normal and become continuous. As the proportion of Xs and Ys changes, the curve would shift in one direction or the other but remain continuous. Now imagine that people are stigmatized if they marry out of their group. If the stigma is harsh enough, one would predict few or no mixed marriages. As numbers increased, the curve would remain discontinuous.

Discontinuity can occur over time, as well. Discriminatory practices can change with extreme rapidity after a long period of stability. Jews had been assimilated in German society prior to Hitler's rise to power; they served in the German military during World War I with enthusiasm and distinction. Their ostracism was extraordinarily swift. Blacks had been outcasts in American society centuries before the civil rights movement; within a generation, the most obvious forms of discrimination had disappeared. These patterns are consistent with signaling models: a behavioral regularity, such as racial discrimination, is self-sustaining and stable over a long period of time against small shocks; then, once a large enough shock occurs, the behavioral pattern vanishes. In contrast, the preference theory predicts that discriminatory behavior would be continuously and normally distributed, just as the consumption of apples is, and that incremental changes in the cost of discrimination would produce incremental changes in discriminatory behavior.

3. People who belong to the in-group often avoid, shun, or even attack others in the in-group who cooperate with members of the out-group. The Serbian militias usually began "cleansing" a village by killing Serbs who were married to Croats or Bosnians (Kuran 1998); whites in the Jim Crow South were especially nasty to whites who treated blacks respectfully (Wyatt-Brown 1982). These patterns are consistent with the signaling theory. Discrimination against people who violate a discriminatory norm occurs because those who deviate from the strategy of the good type reveal themselves to be, or likely to be, bad types; moreover, discrimination against deviants within the in-group can, like discrimination against outsiders, serve as a signal of type. Norm entrepreneurs exploit fears that discriminatory signals will lose their clarity, and push for laws and practices that maintain or enhance the salience of characteristics that provoke discrimination. Hence all the laws, from miscegenation laws in the South to the identification requirements created by the

Nazis, that are designed to make sure that the signal remains clear. The preference theory fails to explain why a person with a preference to avoid members of the out-group would incur the cost of punishing other people in his group who did not share that preference (McAdams 1995, pp. 1039–41).

4. Discriminatory behavior is usually accompanied by theories rationalizing the discriminatory behavior. I mentioned several of these above. These theories respond to two kinds of ambiguity that defeat signaling. The first is the ambiguity about the group to which a potential victim of discrimination belongs. Miscegenation produces people whose features are continuously distributed between the different groups. Theories resolve this ambiguity by turning the inquiry from appearance to ancestry. Race or ethnicity is determined by drops of blood (another fiction), which can be determined by tracing genealogical lines. The second ambiguity concerns the motive of the person who discriminates. Like the altruistic gift-giver, the sadistic racist does not incur costs when engaging in the behavior recognized as the signal. The Nazis invented a racial theory purporting to demonstrate the threat posed by Jews to the purity of Aryan blood, a theory designed to show that discrimination against Jews is not the satisfaction of a private taste. And the signaling theory explains why people would publicly support ridiculous theories that they can easily see through (see Chapter 7). The preference theory does not explain why norm entrepreneurs would invent theories of racial supremacy and why these theories would receive the public support of people who ought to know better.

5. Discriminatory behavior increases rapidly during times of tension and insecurity, a straightforward consequence of the signaling theory. When tensions rise, interaction with strangers becomes more dangerous; thus interaction with partners becomes more important. If discrimination against outsiders is a historic signal of group loyalty, even people who care about the outsiders will be tempted to engage in discrimination as a way of showing their reliability to the group. The wars in the former Yugoslavia tore apart powerful family and community loyalties along ethnic lines. The preference theory cannot explain this behavior. There is no reason to believe that people would be more likely to satisfy a preference for discrimination during times of tension. A similar point can be made about the variety of discriminatory practices across cultures. Discrimination against blacks in the United States, Indians in Africa, Chinese immigrants in Indonesia, Jews in Europe, and so on, should be explained. To say, as the preference theory must, simply that tastes vary across cultures is not to provide an explanation. The signaling theory explains why discrimination often occurs against groups that are perceived as a threat.

6. Discrimination sometimes increases against groups whose skills and en-

dowments have increased. Examples include anti-Semitism and Jewish economic power in Europe, or discrimination against Indians and Indian economic power in Africa. The preference theory implies that as the skills and endowments of outsiders increase, discrimination against them would decline, because discriminating against people with valuable skills and endowments is more costly than discriminating against people without them. The signaling theory is consistent with the increased discrimination, since discrimination can serve as a signal only when it is costly. And because a minority group that prospers when others do poorly will become a target of suspicion among ordinary people, the prosperity of a group may make it a focal point of discrimination.

7. It was mentioned before that discrimination occurs when a person avoids another person on the basis of immutable characteristics such as skin color or ancestry. This claim leaves open the possibility that a person can discriminate against people whose characteristics he shares, as a way of showing the dominant group that he is loyal to it. This happens frequently: one example is the phenomenon of "Jewish anti-Semitism." The signaling theory implies that discrimination against one's own group is likely to occur when outsider discrimination against that group exists, and it explains why members of the outsider group sometimes protect those people (as even the Nazis did). The preference theory cannot explain these patterns. Conversely, some public anti-Semites will have Jewish friends when anti-Semitism is a ubiquitous but not particularly powerful signal (for example, fin de siècle France); and some public advocates of tolerance and ethnic harmony will lack friends outside their ethnic group when the commitment to tolerance is a ubiquitous but not particularly powerful signal (for example, late twentieth century America).

8. A "taste" to avoid minorities is widely considered morally wrong in a way that is different from a taste to avoid goods of various sorts, or even a "taste" to avoid unattractive people. This is a sociological fact that any theory of discriminatory behavior ought to explain. The signaling theory suggests why these attitudes are treated differently: avoiding a person in order to signal one's commitment to others is to use him as a means to an end; avoiding a person simply because one is not attracted to that person is not to do this. This point tracks an influential distinction in moral philosophy. The preference theory conflates behavior that is driven by psychological (even if culturally mediated) factors, such as tastes for food, and behavior that is driven by the imperative to remain in good standing in one's social group.

The Role of the Law

The social cost of discriminatory behavior is obvious, but it is worth mentioning that, unlike the examples of flag burning, self-censorship, and voting, the signal is an injury to a third party who is not a player in the game. The members of the minority are not the "bad types"; those who fail to discriminate against them are the "bad types" (where "good" and "bad" are defined from the perspective of the group). For example, whites in the old South showed their loyalty to each other by discriminating against blacks, and in this game perhaps bad types as well as good pooled in the behavior. Blacks of both types were injured by the discrimination, and indeed played a separate game among themselves in which the bad types (the Uncle Toms) submitted more readily to white racism. Although any rationalization that justifies the view that people who fail to discriminate against a certain group are cheaters will almost certainly conclude that members of that group are cheaters as well, special venom is directed to the insider who breaks ranks and treats outsiders with respect (McAdams 1995, pp. 1039–40; Kuran 1998).

Epstein (1992) argues that discriminatory hiring is a rational governance practice in firms. Epstein assumes, as the signaling theory does, that employers must forge cooperative relationships with employees, and they must facilitate cooperative relationships among employees, because such relationships are value-maximizing when the labor market cannot supply perfect substitutes. Epstein, however, makes the further claim that although market segmentation will occur, discriminatory wages will not; but this claim depends on the assumption that discriminatory behavior is randomly and uniformly distributed. Now it might be true that discriminatory *preferences* are randomly distributed, but the signaling model suggests that discriminatory *behavior* is not. If whites receive reputational as well as intrinsic returns from not hiring blacks, even whites who like associating with blacks will refuse to hire them or work with them. Thus, Epstein's optimism depends on the premise that discrimination no longer remains a powerful signal of loyalty today, the way it was forty years ago in the South.

Epstein uses his theory to criticize anti-discrimination laws, but the economic case against anti-discrimination laws is actually quite weak. Because Epstein believes that discrimination is a rational governance structure, he can conclude that laws that interfere with it will reduce profits, wages, and employment, and increase prices paid by consumers. But if white employers discriminate against blacks for reputational purposes, then a law that prohibits such discrimination could make all the employers better off.

It seems plausible that this is the case, but one should not underestimate the risks of government efforts to transform signals, as the prior chapter argued. As long as the felt need to identify cheaters remains, norm entrepreneurs will search out new signals.[4] Striking examples of the resulting symbol transformation come from the history of anti-Semitism in Europe. In Spain in the fourteenth century and again in Germany and elsewhere in the nineteenth century, discrimination on the basis of Jewish religiosity lost its effectiveness for causing separation. In Spain the reason was that persecution drove Jews to emigrate or convert to Christianity; in Germany the reason was that theological grounds for discrimination lost their plausibility as the influence of religion declined generally. But the great need in both countries for national unity provoked a demand for a method for distinguishing the loyal from the disloyal, a new theory of discrimination. The norm entrepreneurs of the time—Catholic officials in Spain, nationalist ideologues in Germany—invented such a theory. The new theory was based on race, not religion. People who used to practice Jewish rituals or whose parents practiced Jewish rituals were classified as racial Jews, even though in Spain—as sophisticated contemporaries understood—just about everyone had Jewish blood! The norm entrepreneurs dealt with the problems created by demographic and theological trends by redefining the body of people that could be the subject of discrimination, and their ideas were enthusiastically received by people searching for a way to signal their loyalties.[5] The theories succeeded because the entrepreneurs seized on one strand of a long tradition of discrimination and tolerance. Racial ideas were in the air; discrimination against Jews had a long history. The entrepreneurs succeeded by focusing attention on these themes and diverting attention from others, creating a focal point around which people could coordinate.[6] (By contrast, in nineteenth century Italy the obvious barrier to unification was not the Jewish community but the military power of foreign occupiers; in the United States the obvious barrier to unification was the problem of slavery and the differing economic interests of North and South. So anti-Semitism was not as salient and hence not as well-received.)

There is a further point about anti-discrimination law. If employers freely discriminated on account of race, and as a result blacks earned lower wages than equally productive whites (a result that Epstein quarrels with), then blacks would have a smaller incentive to invest in human capital than whites do. In this way false beliefs about blacks can become self-fulfilling prophecies, and therefore legal intervention to disrupt this cycle is justified (Arrow 1973). Coate and Loury (1993) show that affirmative action may be justified for a similar reason. If in equilibrium employers believe that blacks have inferior skills and so refuse to hire and promote them, they will never observe their skills and never have a chance to correct their beliefs if they are false; therefore, again, blacks have little incentive to invest in human capital. But if affirmative action is instituted, blacks will anticipate that employers will hire and promote them, so blacks will invest in skills, and employers will observe that blacks have skills and correct their false beliefs to the contrary. The landscape is hardly clear, however. Coate and Loury's model produces another equilibrium in which blacks rationally underinvest in skills because affirmative action enables them to obtain good jobs without having sufficiently developed skills. Until evidence on these issues has been gathered, an economic analysis that properly takes into account the role of belief and reputation in the labor market will not support either opponents or defenders of affirmative action.[7]

Nationalism and Nationalist Myths

Discussions of nationalism emphasize the paradox that members of a "nation" feel a profound connection with thousands or millions of people whom they have never met and with whom they share mostly imaginary attributes, while hating with great ferocity those who lack these same, mostly imaginary attributes. There are two points that are worth emphasizing. First, most nations define themselves using criteria that are incorrect and arbitrary. Usually, the nation claims that its people share common ancestors: everyone belongs to the same ethnicity and derives from the same tribe. Yet historical migration and intermarriage make nonsense of these claims. Or nations claim a common language or culture, whereas in fact one language is declared the national language, and all the other languages within the nation's territorial borders are declared to be dialects of the national language—with the line between insider languages and outsider languages set by a real or aspirational border. Similarly, the culture of the most powerful region is attributed to people in other regions who are brought within the territorial boundaries of the nation. Second, it is hard to understand how these slight differences, even if they were real, would justify violence and conquest. Indeed, justificatory national myths of humiliation and revenge are imposed on all people who are said to belong to the nation, so that, again, whether a living person is said to suffer the humiliation of a military defeat incurred centuries ago depends on whether that person's ancestors happened to live on one side or the other of a border that was later drawn on the basis of independent political forces.

These ideas are captured in the phrases, "imagined communities" (B. Anderson 1983) and "invented traditions" (Hobsbawm and Ranger 1983). A nation is an imagined community, not a real community, since there is little or no contact between its far-flung members, and not much in common. To justify the claim of a national bond, traditions are invented that give the members of the nation something in common—a common history. The problem with the phrase "invented traditions" is that it implies agency when traditions usually arise spontaneously. Although, as Hobsbawm and others have shown, national governments and private norm entrepreneurs have played an important role in inventing traditions, they had to have a receptive audience. "Spontaneous tradition" preserves the paradoxical flavor of nationalism without implying that influential individuals necessarily play a role in the creation of tradition. The literature on nationalism focuses on how nationalism is possible, and on the closely related question of the origin of nationalism. The latter question complicates the puzzle. Most historians believe that nationalist movements are a recent phenomenon. There were no national movements, and nothing like modern nationalist sentiment, prior to the nineteenth century. Yet the two great wars of the twentieth century had a strongly nationalist character, and after the brief hiatus brought on by the ideological conflict of the Cold War, nationalism is reasserting itself around the world.

The signaling model sheds light on nationalist movements.[8] Nationalism (and related movements, such as isolationism and protectionism) arises during times of worldwide or regional insecurity, when the trustworthiness of one's cooperative partners becomes increasingly important. Discriminatory attitudes and practices against those who are defined as outside the nation become signals. Examples are numerous, but one can point to the resurgence of nationalism after the collapse of the central governments in Yugoslavia and the Soviet Union; nationalism in Germany in response to its insecurity prior to unification in the nineteenth century and again after World War I; the growth of Zionism in response to the sudden outbreak of pogroms in the nineteenth century after a long period of relative tolerance of Jews (Klier 1997); and recent nationalist trends in France and other European countries in response to economic stagnation.

When everyone in a "nation" believes that everyone else in that nation is trustworthy, or at least more trustworthy than outsiders, then even strangers within that nation can cooperate. Compare a kin-based system, in which only a small group of blood or marital relations can cooperate with each other. A nation of people who trust each other, even if in an attenuated way, is more powerful than a tribe or clan, everything else being equal, simply because the nation is larger. A nation is more powerful than a state or empire that uses force to join together separate tribes or clans, because the state is weakened by antagonisms among its smaller units. The innovation in nineteenth century politics was due to the emergence of a real if attenuated solidarity between larger numbers of people than ever before, enabled by advances in communication (the telegraph, printing) and transportation (the railroad). These technological advances, which explain why nationalism arose in the nineteenth century and not earlier, could not produce solidarity beyond the limits imposed by language and distance. This is why linguistic uniformity and territorial coherence are dominant themes in nationalist rhetoric.

The question remains, How does a group of people form a nation? Why do some groups become nations rather than others? Are there limits to the size of a nation?

Nationalism exists because a highly successful signal is that of shunning people who are outside the nation. "Nation" is not exogenous: what counts as a nation, just as what counts as a race, depends on the economics of signaling. The person to be shunned must have exactly the right characteristics. He must have characteristics different from those of the people with whom one would like to cooperate, and these characteristics must be easy to determine. If the people with land, wealth, valuable skills, or political power in a particular geographic region have a certain color skin, or physical build, then people within that group and perhaps also those on the margin will signal their loyalty to the insiders by shunning those who have differently colored skin or a different physical build. Hence the frequent overlap between nationality and ethnicity or race. But the group of desirable cooperative partners may *not* have physical characteristics that are systematically different from the characteristics of undesirable partners, in which case shunning people without the insiders' characteristics cannot serve as a signal. But one can also relatively easily observe linguistic and cultural competencies. If the people with land, wealth, valuable skills, or political power in a particular geographic region speak the same language or share a culture, then people within that group and perhaps also those on the margin will signal their loyalty to the insiders by shunning those who do not speak that language or enjoy that culture. People signal their commitment to insiders by investing in linguistic and cultural competence, a costly and observable action. Shunning those who do not speak one's language or culture is an attractive strategy, because a shared language and culture facilitate cooperation by reducing the risk of miscommunication (Gellner 1987, p. 15).

Still, it is not always the case that the group of people with land, wealth, valuable skills, and political power share a language and culture. If the people with land, wealth, valuable skills, or political power in a particular geographic region share a real or mythical (but widely believed) ancestry or have ancestors who had some real or mythical (but widely believed) connection (such as defeat in an ancient battle), then people within that group and perhaps also those on the margin will signal their loyalty to the insiders by shunning those who cannot plausibly make similar ancestral claims. Hence the frequent role of history and myth in the definition of nationality. Because the definition of the nation is always in these ways a matter of expedience and accident, it is not surprising that different nations are defined in different ways, and that all of the elements mentioned, and many more besides, usually play a role in national conceptions.

Discrimination is a valuable signal for other reasons as well. Discrimination is always the poor man's signal: one does not need any assets in order to issue this signal, unlike the case of gift-giving or clothing fashion, and yet it still is costly because one forgoes employment, commercial, and social opportunities. Moreover, discrimination, unlike, say, gift-giving, forces the injured group to adopt a symmetrical norm. When ethnic Serbs begin to discriminate against Croats, reciprocal discrimination is a salient response among the Croats. This is a characteristic of signals that have strong third party effects: they draw the third party into a game that otherwise is being played between the good types and the bad types of the main group. Finally, discrimination is highly flexible, because the target can, within constraints, be defined as those who lack the characteristics of those who are desirable. The convenience of discrimination against those who lack cultural competence almost compels the idiosyncracy of culture: if one could deduce a person's cultural beliefs by applying an algorithm in one's head, then one could imitate the native easily, and cultural competence could not serve a signaling function (compare Gellner 1992, p. 149).

Nationalist movements emerge because discrimination on the basis of nationality becomes a signal of loyalty. A nation forms from a collection of personal characteristics when these characteristics are readily observed, enough people have them to obtain more power relative to other possible groupings (for example, by class or religion or race), circumstances grant economic and political advantages to people who are able to engage in large-scale cooperation, and—this is crucial—people manage to develop the belief that those with the appropriate characteristics are more reliable cooperators than those without them, a belief that becomes self-fulfilling as soon as enough people hold it (but is false before then). People may even recognize that the beliefs are false, but they cannot openly say that they are false, for that would be to fail to send the signal of loyalty, and out of the embarrassed silence of the first generation are forged the iron convictions of the second.[9] This returns us to the imaginary or invented element that plays such an important role in nationalism, and brings us to the norm entrepreneur.

Norm entrepreneurs, political entrepreneurs, and government officials earn returns by stirring up ethnic hatred, and they do so by engaging in signaling themselves. They appeal to historical humiliations, and they invent new ones. They invent rationalizations for discrimination, prejudice, and war. This is not to say that they are invariably successful. The clumsy efforts of East Germany to create a myth that would unify its citizens failed dramatically (Fullbrook 1997). And the ability of entrepreneurs to create myths and traditions is constrained by the material at hand: whether, for example, there is a glorious antiquity or humiliating defeat to appeal to (A. Smith 1997). Nationalism is also limited by this tendency toward exaggeration and myth-making, which brings nations into conflict with each other, and exposes any nation to counter-myths created by subordinate ethnic groups within its borders. But the fact remains that charismatic people, or people who already have the attention of the crowd, can stir up nationalist feelings by persuading a core group of receptive ears, then letting the dynamics of signaling take over once a certain threshold of belief is met.

This model of nationalism is just a sketch, but its usefulness for understanding nationalism can be seen in its focus on individual incentives. It shows how the individual's concern with his reputation can lead to behavioral cascades that are observable in the behavior of millions. For all its apparent irrationality—its reliance on myths, its tendency to war and self-destruction—nationalism can be explained as the outcome of individual rational action.

9

Contract Law and Commercial Behavior

The model of cooperation described in Chapter 2 focuses on *nonlegal* mechanisms of cooperation. It suggests that in a variety of circumstances people can cooperate without the intervention of the legal system. But if people can cooperate in the absence of a legal system, then what does the legal system do? Does it interfere with their ability to cooperate, or does it enhance their ability to cooperate, or does it have some other effect?

Chapters 4 through 8 suggested that the legal system will sometimes enhance cooperation and sometimes interfere with it. When powerful nonlegal mechanisms of regulation exist, the state is tempted to channel them. The state subsidizes charitable organizations, in order to stimulate public giving, and protects private transfers of wealth but forbids destructive dowry competitions. It uses shaming penalties to stimulate the mob against criminals, then conceals punishments and criminal records in order to restrain the mob. It formalizes the marriage vow and defers to household discipline, while regulating the dissolution of marriage in minute detail. It uses propaganda to stimulate patriotic and nationalist sentiments, but punishes those who discriminate against traditional racial or ethnic antagonists who are included in the "nation"—in this way government officials seem to try to define the insiders and the outsiders in the "optimal" way (whatever that is), then hope that the dynamics of signaling and cooperation will take over, creating mass enthusiasm that makes the standard tools of state coercion unnecessary.

This chapter carries on this theme but approaches the problem in a different way. Now I focus on commercial behavior, a topic glanced at in Chapters 1, 2, and 4. Those chapters noted that merchants must cooperate with each other in order to make profits, but cooperation is hampered as always by incentives to cheat. The long history of commercial behavior is powerful evidence that merchants can overcome these incentives much of the time—enough of the time, anyway, to be able to prosper—despite the absence of legal intervention. Examples include the Lombard and Jewish bankers in the early modern period, the Maghribi traders, the Genoese and the Venetians, ethnic Chinese merchants in foreign countries, Korean and other immigrant groups in the United States, and the successful exploitation of common pools by local groups.[1] In all of these cases, merchants cooperate and prosper in a lawless environment at the international level, or even in a hostile local legal environment. So one might ask, Why is a state necessary at all for commercial cooperation?

A common answer is that although relatively small and homogenous groups of people can cooperate, citizens of a populous state cannot. But this answer cannot be the whole story. Most people are able to cooperate without resorting to the threat of legal sanctions. In ordinary life, people constantly make and keep promises; and legal retaliation for cheating is never an option because the cost of invoking the law exceeds the amount at stake. The common wisdom might be revised, then, to hold that nonlegal cooperation occurs among people in communities, where information flows freely and reputations are known, but not among strangers.

Even so revised, however, this view is unsatisfactory. What are these contracts among strangers? When a consumer purchases a stereo at a retail outlet, the consumer and the "store"—whether we mean the salesclerk, the manager, or the shareholders—are strangers, but it is rare for a consumer to sue the store if the stereo is broken. No rational consumer would sue a store over an object worth a few hundred dollars, when the lawsuit would cost the consumer thousands of dollars. But a lawsuit is rarely an issue, anyway; most retailers offer warranties and honor them because they fear damage to their reputation. And if the consumer does not honor a promise to pay for the stereo, the retailer might sue the consumer, but more likely it will report him to a credit agency that will record the default on the consumer's credit report. So the retailer and the consumer are not really strangers, or if they are, then they embarrass the claim that nonlegal cooperation does not occur among strangers.

Perhaps, then, the "contracts among strangers" refer to arm's-length sales among merchants. Even here, however, reputation and other nonlegal mechanisms play an important role. Most merchants belong to trade associations, clubs, and other organizations, which enable them to meet each other and exchange information. A large company may have thousands of employees, but all the employees with major responsibilities will attend conventions where they meet their counterparts in other firms. So what appears to be an arm's-length contract between two anonymous firms is often the result of negotiations between two friends who belong to the same social club or sit on the board of the same charitable organization. An enormous amount of business activity consists of making contacts, or "networking," and what does this mean if not revealing information about oneself to others, and obtaining information about them in return? Once one has enough information about someone, then one might trust him enough to do business. Without such information, the paraphernalia of the legal system—the judges, the courthouses, the clerks, the accretion of precedent—are cold comfort.

So when contracts are small, people do not sue each other because it is not worthwhile. When contracts are large, people do not sue each other because they depend on reputation. But if this is so, what is the role of the law? Put differently, if the law were adequate for regulating relations among strangers, then why wouldn't people rely on the law rather than spending so much time and effort worrying about reputation?

Commercial Behavior

The traditional paradigm of contractual behavior generally assumes that people make contracts because only legal sanctions will deter a party from cheating on the contract when it is profitable to do so. If each party expected the other to cheat under such conditions, parties would not enter a contract in the first place. The value-maximizing court enforces contracts in such a way that maximizes the ex ante value of the contract, which usually means allocating obligations in a way that places the risk of any contingency on the party that can most cheaply bear it and that gives them proper incentives to breach, invest, and engage in related behavior.

This paradigm misdescribes modern commercial practices in many ways. I have mentioned several. Parties to a contract are rarely strangers to each other. In almost all contracts, one party or both parties care deeply about their reputations. In ordinary commercial contracts between merchants, both merchants expect to do business with each other in the future, or at least with other merchants who are likely to learn about the behavior of the parties. Employers and workers understand that employment contracts cannot provide for all the behavior that will be required on each side. Workers behave properly in order to obtain bonuses and promotions and in order to avoid being penalized or fired. Employers behave properly in order to maintain the loyalty of their workers and to attract workers entering the market. Even something as transitory as a stock transaction is constrained by nonlegal sanctions. The buyer and seller in the secondary market do not deal with each other. They both deal with a middleman, the broker, who takes pains to develop a reputation for honesty, and who usually is employed by a firm with a brand name, built up over a number of years.

Other kinds of behavior are hard to explain if one assumes the traditional paradigm. Contracting parties are often friends. A book publisher might take a client out to lunch or dinner. Purchasing agents take suppliers to baseball games, plays and movies, even to strip-tease joints (Meredith 1997). Business deals are everywhere forged in bars, restaurants, and private clubs. Business is almost always conducted in a highly social manner. First, participants talk about sports; then, about their families; and only *then,* perhaps when the dinner or golf game is almost over, do they shake hands on the deal.

In the cotton industry, "Merchants take mill buyers on hunting trips just like in any other business . . . In the process, relationships . . . develop[]. Over time a buyer gets the idea that he wants to deal with me not just because of our business relationship, but also because of our personal relationship. So you tell me, when you want to do business who will you call, the guy you like or the guy you don't like?" (Bernstein 1999, p. 16 (quoting merchant, brackets and ellipsis in original)). A major trade association "has sponsored the local debutante ball, an annual civic cotton carnival, golf tournaments, a Cotton Wives Club [sic], a well-known domino tournament, and numerous other civic events. To this day it continues to encourage social interaction among its members and their families by making its annual conventions family events" (Bernstein 1999, pp. 20–21) Many businesses, trade associations, and other industry groups sponsor social and family events in order to enhance relationships among their employees or members.

As everyone knows, business is frequently a family matter. The patriarch hires and grooms his children, then transfers ownership to them while he is still alive or upon his death. A husband and wife own and manage a laundry or grocery store, where the children participate as employees. This phenomenon is easy enough to understand. Business based on ethnic relations is harder to understand. People who belong to an ethnic group may know as little about each other as they know about outsiders, yet they are drawn together in common enterprises. A vast number of sociological case studies describe business relationships that are based on ethnic ties. Only ethnic Koreans are invited to participate in the kye, a kind of rotating credit group; and only ethnic Mexicans are invited to participate in the cundida or tanda, their version of the rotating credit group (E. Posner 1996c, pp. 168–71). Many small businesses, particularly in ethnic parts of cities, will hire only co-ethnics, even though these co-ethnics may be no better known to the owners than applicants who do not belong to the ethnic group.

If the reliance on ethnicity in business is puzzling, even more puzzling is the reliance on fictive kinship, where unrelated people treat each other as though they were kin. Commercial "brotherhoods" (or confraternities) from the Middle Ages are familiar, and they survive in the fraternal ideology of labor unions. Here is a modern example of fictive family relations in business, described in a strikingly vulgar (and somewhat confused) way:

> actually what [franchising] is, is a wedding. Lots of music, lots of flowers, money exchanging hands and lots of kisses. The couple is from the best of two worlds; one of the partners is experienced, with plenty of good know-how, with a proven system; and the other partner is a virgin, who hopefully has never been in business before. The vows they exchange are almost the same as you exchanged when you married your wife. The virgin bride must have a burning desire to be "his" own boss and to run "his" own business. (Hadfield 1990, p. 965, quoting Coomer 1970)

Finally, the traditional model puts too big a burden on courts. Many scholars acknowledge that courts cannot determine obligations in long-term, "relational" contracts, in which many terms are left out, but appear to believe that courts can determine obligations in shorter, one-shot deals. This latter claim is difficult to confirm or deny, but let me mention two reasons why it is unlikely to be true. First, although the number of unpredictable contingencies that can change the value of a long-term relationship is no doubt huge, the number of unpredictable contingencies that can change the value of one-shot deals is also huge. The fact that the number of contingencies is overwhelming in the first case does not imply that the number of contingencies can be handled in the second case. The relatively discrete sale of a house extends over months during which any number of things can happen, only a small fraction of which can adequately be treated in the contract. Short-term contracts almost always have tails stretching indefinitely in the future. A buyer might sue the seller for a defect in goods discovered months or years after delivery of the product, a suit that requires the court to determine whether intervening contingencies are relevant in the determination of obligations.

Second, courts have trouble understanding the simplest of business relationships. This is not surprising. Judges must be generalists, but usually they have narrow backgrounds in a particular field of the law, and they often owe their positions to political connections rather than merit. Their frequent failure to understand transactions is well-documented. One survey of cases involving consumer credit, for example, showed that the judges did not even understand the concept of present value (Allen and Staaf 1982). The judges struck down contracts using the unconscionability doctrine because the credit price was higher than the cash price, but they did not take account of risk and of the time value of money, which were not out of the ordinary. Even when judges do not misunderstand basic ideas, one must take their interpretation of facts on faith. Judges' reasoning can be evaluated only against the canned facts described in the opinion, which themselves are the result of a fact-finding process that does not inspire confidence. Given the varying sophistication of trial judges, lawyers, and juries; the accidents of discovery; the unpredictability of witnesses; and the inescapable vagueness of the law, parties could be forgiven for believing that the chance of winning a breach of contract suit is pretty much random. Indeed, skepticism about the quality of judicial decision-making is reflected in many legal doctrines, including the business judgment rule in corporate law, which restrains courts from second-guessing managers and directors, and the many contract law doctrines that restrain courts from second-guessing parties to contracts.

These observations suggest the following possibility. Courts are not good at deterring opportunistic behavior in contractual relationships, but parties are. This is why so much contractual behavior depends on reputation, ethnic and family connections, and other elements of nonlegal regulation, and not on detailed and carefully written contracts enforced by disinterested courts. The next section analyzes this hypothesis more formally, and suggests an answer to the question, Why, if they cannot rely on courts to enforce contracts properly, do people so frequently take pains to ensure that their contracts are legally enforceable?

A Model of Commercial Behavior

Several scholars have noted that contracts are often radically incomplete regarding the obligations of the parties (Macneil 1978, Goetz and Scott 1981, Hadfield 1990, Schwartz 1992, 1998). Most of these commentators, however, assert that judges can complete the contracts ex post by supplying value-maximizing terms or by enforcing business norms.

Only two commentators have confronted the possibility that judges do not have enough information to allow them to supply appropriate terms when contracts are incomplete. In a mostly positive analysis of judicial response to incomplete contracts, Schwartz (1992) argues that courts adopt either an "active" or "passive" strategy toward contract disputes. The active strategy is the traditional one of filling gaps in incomplete contracts by constructing the ex ante optimal contract terms. The passive strategy involves literalistic enforcement of contract terms even though everyone knows that the contingency that gave rise to the dispute was not contemplated by the parties when they entered the contract, and so the terms are unlikely to describe optimal obligations in light of that contingency. Turning from positive to normative, Schwartz argues that courts are competent enough to fill gaps in relatively complete contracts, and they should, for the traditional reasons that appear in the literature. More tentatively, Schwartz suggests that the literalistic approach is suitable for highly incomplete contracts, because gap-filling will fail, and literalistic interpretation at least enables the parties to rely on a set of fixed contractual obligations when renegotiating in light of changed circumstances (see also Schwartz 1998). Schwartz's argument, however, depends on the assumptions that parties can anticipate contingencies with adequate precision, that they can draft sufficiently detailed terms in light of these anticipated contingencies, and that courts can enforce these terms accurately. These assumptions may not be correct and I will relax them.

Hadfield (1994) argues that even incompetent judges should fill gaps in incomplete contracts. If judges' decisions are inaccurate but better than random, then their decisions will sometimes punish opportunistic breach and sometimes not. Anticipating this result, the potential breacher will sometimes but not always be deterred from engaging in opportunism, which is a better result than if the potential breacher is never deterred. The existence of judicial error increases the value of the contract to whichever party is in a better position to breach and reduces the value of the contract to the other party, but the ex post beneficiary must compensate the ex post loser with a lower or higher ex ante contract price. If the judges' decisions are completely random, no deterrence will occur, but, as before, an ex ante transfer will compensate the ex post loser, and so the randomness of the judges' decisions will not affect contracting compared to a regime in which no enforcement occurs.[2]

The problem with Hadfield's argument is that if one assumes radical judicial incompetence, as I do, the system of contract law serves no purpose. Although, as she shows, parties do no worse by entering legally enforceable contracts than by not entering such contracts, they do not do any better, either.

There is an alternative model for understanding the role of courts when contracts are incomplete. The claim underlying the model is that even if courts cannot determine who breached a contract, or whether a contract has been breached, they can deter opportunistic behavior. This claim might sound implausible, but the key to it is that parties choose when they want to use courts and when they do not, so even an uncomprehending court can serve a purpose as long as it allows itself to be manipulated by the parties.

The model is the cooperation model from Chapter 2, with its emphasis on repeat play, on the one hand, and signaling, on the other. The analysis should by now be familiar, so rather than redescribe it, I will show how the oddities of commercial behavior described above illustrate its various aspects.

The dinners, lunches, parties, clubs, gifts, and other apparent epiphenomena of commercial interaction are signals. These activities are costly, and although people may enjoy these activities, there is little reason to believe that they engage in them just because they enjoy them. Most people would rather go to dinner with their (real) friends than with their business associates; most people taken out to dinner with their business associates would prefer the money value of the meal. But cash gifts cannot serve as a signal because cash is fungible. If X gives cash to Y, then X cannot know whether Y is attracted to X for X's business or for X's money. In bilateral signaling, both sides must waste time, effort, and money.

A strategy can succeed only if the receiver of the signal knows or can accurately estimate the cost structure of the signal. The receiver must consider at least two possibilities. The first is that the sender of the signal has idiosyncratic tastes—or, at the extreme, is crazy. The second is that the sender has a low discount rate—is a good type who can incur present costs because he values future returns. Although any action with the right cost structure can serve as a signal, people are more familiar with the cost structures of some actions than others and thus can more easily interpret them. Burning money and running around in circles are hard to evaluate because most people do not burn money or run around in circles; sending a holiday card or taking someone out to dinner, by contrast, are easy to evaluate. Thus, those who want to reveal their type engage in costly behavior that has proven in the past to be a useful signal for other people. This is why so much signaling behavior has a ritualistic quality. People imitate what other people are doing without thinking much about the meaning of the behavior. This theory also explains why in different business cultures, arbitrarily different behaviors become entrenched. In one place people bowl, in another place they play baci, in a third they attend strip-tease shows.

Family-, friend- and kin-related signaling is pervasive because parties minimize ambiguities about their signals by drawing analogies to activities in which relations are better understood. Entering a new contract is likened to a marriage, so that seriousness of the relationship is understood. Time and expense are incurred in the contract ceremony for the same reason that they are incurred in a marriage ceremony: to signal discount rate. This conclusion also holds for lengthy "courtships," where costs are incurred by forgoing opportunities to pursue relationships with other people or businesses. The rituals of family life—the cycles of gift-giving, entertainment, and the formalities of deference, respect, and equality—are carried over into business life.

There are independent but complementary explanations for the pervasiveness of family relations in business life. The manager of the firm has more information about family members than about strangers, including information about discount rates and skills. And the manager of the firm might have more control over co-ethnics than outsiders. If a worker belongs to an ethnic group that suffers discrimination at the hands of a dominant majority of outsiders, if that worker therefore can obtain well-paying jobs only from co-ethnics, and if reputations travel rapidly within the ethnic community, then the employer can discipline the worker by threatening to fire him and broadcasting his incompetence to co-ethnic employers. Ethnic solidarity, produced by the hostility of outsiders, give insiders an effective way to ensure cooperation. These phenomena are independent of the signaling argument, and less interesting, so I will put them aside.

The cooperation model can explain why people do not break promises under a variety of circumstances, but it cannot be a theory of contract law. It shows why people keep their promises even when the legal system is expensive or unavailable, but it does not show why people often take great care to enter a contract that is legally binding. I turn now to this question.

Assume that a legal system exists but that courts are unable to determine whether a party to a contract has broken a promise. To be more precise, assume that a person can ask a court to give him a remedy for breach of contract, but that the court is so prone to error that its decisions are random as to liability, with damages being represented by an unbiased distribution around the amount at stake. Suppose, for example, that X cheats Y, with the result that 50 is transferred to X at a cost of 100 to Y. Y sues for 100. The court holds with 50 percent probability in favor of X and with 50 percent probability in favor of Y. If the court holds for Y, damages will be normally distributed around 100. If the court holds for X, damages will be normally distributed around 0 (with negative damages interpreted to mean that the court holds in favor of X on a counterclaim for a positive amount). The assumption can be understood as one of judicial incompetence or error-proneness but it is equivalent to the assumption that whatever the intelligence and sophistication of judges, parties cannot anticipate and contract about any contingencies in a sufficiently fine-grained way to provide guidance to such judges.

I also assume that even though a court cannot determine whether a promise has been performed or breached, it can determine whether the parties have intended to enter a contract. I will defend this assumption below.

Two parties, call them Seller and Buyer, enter a contract whose ex ante value to each exceeds the value of the next best opportunity. The contractual relationship has the form of a repeated prisoner's dilemma, so that the ex ante value of the contract is realized only if each party can overcome its incentive to defect. A party always observes a defection, and can retaliate by refusing to cooperate in later rounds. A court cannot observe a defection. In the standard vocabulary, defections are observable but not verifiable.

Now suppose that two kinds of events can occur that will affect the parties' incentives to defect: a small change in prices and a large change in prices. If a small change in prices occurs—whether to the advantage of a buyer or a seller—neither party defects, because the gain is less than the discounted value of continuation of the relationship. If a large-scale change occurs, then the party favored by the change will defect (under current assumptions), because the gain exceeds the discounted value of continuation of the relationship. The role of the law, I claim, is to deter such defections caused by large-scale price changes.

But how can the law deter defections if, by assumption, courts cannot verify the underlying event? If the Buyer refuses to take delivery, the court does not know whether the Buyer is defecting, or is retaliating for an earlier defection by the Seller with respect to quality.

Suppose, however, that if one party sues another, both parties expect to incur a substantial cost, C. The reason is that litigation is a negative sum game: one party must pay the other party damages or neither pays the other damages, but in either case both parties must invest a great deal of time and money in lawyers and litigation. Suppose, moreover, that the payoff from defection when the other person cooperates, D, is less than C. Then, if both parties choose simultaneously whether or not to invest in the relationship-specific assets after having entered a contract, mutual cooperation and mutual defection are equilibriums. However, mutual cooperation is more plausible, because in what is now a simple coordination game the player's interests converge: each receives a higher payoff from mutual cooperation than from mutual defection.[3]

This argument rests on the assumption that each party can credibly threaten to sue the other if that person defects. One might doubt that this assumption is valid. Imagine that Buyer cooperates and Seller defects, then Buyer must decide whether or not to sue Seller. If Buyer does not sue, he is stuck with the low sucker payoff, S; but if he does sue he receives S - C, which is even lower. So Buyer would never sue; therefore, Buyer's threat to sue is not credible; Seller would not be deterred from cheating; Buyer must expect Seller to defect; and Buyer would not enter the contract in the first place.

To avoid this outcome, one must make another assumption. Buyer (Seller also) cares about having a reputation among third parties for being tough: if anyone cheats him, then he will retaliate by suing. Buyer wants this reputation, because if sellers believe it, they will not cheat him. This reputation is credible as long as Buyer actually sues anyone who cheats him. Now it pays Buyer to sue anyone who cheats him, as long as the short-term loss, C, is offset by the long-term gains resulting from the future contracts, in which future sellers do not cheat Buyer; gains that are possible only if Buyer establishes a reputation for toughness by suing anyone who cheats him. If Buyer's discount rate is sufficiently low, Buyer will sue any seller that cheats him. This costly lawsuit serves as a signal to future sellers that Buyer is a reliable type.

Although C must be high enough to deter Seller from cheating, C cannot be higher than Buyer's discounted gain from having a good reputation. (Recall that we assume that C is the same for Buyer and Seller.) If C is too high, then Buyer would not sue Seller after Seller cheats, and, knowing this, Seller would cheat rather than cooperate. It follows that if D, the defection payoff, is very high, Buyer cannot deter Seller from cheating. Seller knows that Buyer would have to impose a very high cost on her, but because an expensive lawsuit would hurt Buyer as well, Buyer would rather avoid the lawsuit than establish a reputation for toughness. The availability of incompetent courts, then, enables more cooperation than would exist in their absence, but clearly—and intuitively—does not make full cooperation possible.

Small-scale changes in price do not result in breach, because the promisor values the discounted future payoffs from the relationship with the promisee more than the one-time gain from cheating. Large-scale changes in price do not result in breach, as long as (in our example) Seller's one-time gain from cheating is less than the Buyer's discount future payoffs from relationships with *all* future transactors. Clearly, future payoffs from all transactors will be greater than future payoffs from a single transactor: that is what drives the conclusion that deterrence is improved. But the explanation is more subtle than it might appear. Breaches based on small changes do not occur because of the credible threat to retaliate in a two-party game in which both parties receive substantial gains from continuing the relationship. Breaches based on large changes are *not* deterred by retaliation by third parties, nor by the threat of judicial enforcement—the cost of information makes such mechanisms impossible. Rather, breaches based on large changes are deterred because courts, by imposing costs on the victim as well as wrongdoer, enhance the credibility of the victim's complaint, and publicize it as well.

An important question is how C is determined. Notice that C is not the same as, say, expectation damages. When Buyer sues, he can spend as much on litigation as he wants to. How much will he spend? He will not settle with Seller, because then he will not obtain a reputation for toughness. He must spend enough that observers see that he is willing to impose enough costs on Seller so that Seller gains nothing from cheating. If we make the plausible assumption that one's chances of winning a lawsuit increase with the amount of money one spends on litigation, then Buyer can force Seller to incur litigation costs simply by incurring litigation costs himself. To prevent Buyer from winning, Seller will try to match Buyer, thus keeping the odds even. So Buyer will spend C, forcing Seller to spend C as well.

Now return to the assumption that the parties can choose whether or not their relationship will be subject to legal intervention. This assumption is necessary to the argument. If a party could involuntarily be subject to contractual liability, or if parties could not choose to be contractually liable, then people would have an incentive to make fraudulent claims that strangers have entered contracts with them, as a way of extracting value from the strangers. In addition, a party always has an interest in binding the other party but not himself. Seller might be willing to enter a legally binding contract with Buyer; but Seller would prefer a legal relationship in which she (the Seller) is not bound and Buyer is. So if Buyer cheats, Seller can credibly threaten to sue Buyer. But if Seller spots a good opportunity to cheat, she can cheat without fearing a lawsuit from the Buyer. To prevent these outcomes, our incompetent court must not be *too* incompetent: it must be able to distinguish a legally binding promise from a non-binding promise. Courts are too incompetent to fill gaps in contracts, but not too incompetent to determine whether the parties intended to make legal remedies available to each other.

Is this a plausible claim? Not even the parties know what the optimal terms are ex ante, so they cannot embody them in a complete contract. The optimal terms emerge through interaction but they are never obvious; they must be interpreted from the behavior. This is likely beyond the ability of any court. But parties probably do know ex ante whether they want a court to be involved if a dispute arises. They want a court to be involved simply if they need to have available to them the ability to threaten to impose huge costs on the other party, when the potential gains from cheating by the other party are very high. Interpretive difficulties for courts can be avoided as long as the law stipulates forms to which the parties must adhere in order to signal to courts their desire for legal enforcement, and the parties know these forms in advance. If the law says that a seal is a signal of a desire for legal enforcement, the parties can unambiguously signal their desire for legal enforcement by attaching a seal to their written contract.

A crisp example of this phenomenon—confidence in courts' ability to evaluate the use of form but not to determine obligations and evaluate performance of them—comes from family law. Most people believe that courts have little ability to evaluate an ongoing marital relationship. The evolving obligations are too complex for an outsider to understand. Although courts and other agencies intervene more in family relationships now than they did in the past, it remains true that families enjoy a great deal of autonomy. But—now, as in the past—parties are required to commit themselves or not to legal enforcement by either getting married or not getting married. It is thus assumed that courts *can* determine whether two people are married. We see here the sharp distinction between the assumption that courts can understand the use of form to signal an intention to enter a legal relationship and the assumption that they nevertheless cannot resolve the disputes that arise within that relationship. Although commentators sometimes say that marriage is a kind of contract, the reverse is the correct statement. A contract is a kind of marriage.[4]

The model of contract law that I have described does not put great demands on the courts. It is as though two parties to a relationship agreed that if they had a dispute, both parties would have a finger chopped off by a government agent. Neither party cheats, because he believes that the other would retaliate by invoking his right to have the mutual sanction imposed. The cheated party will credibly retaliate, because otherwise he risks obtaining a reputation as a softy, in which case he will be unable to avoid being cheated the next time he plays this game. The government agent's role is just to chop off fingers of both parties if one party complains. The government is like a parent, who punishes both children who are fighting rather than only the child who started the dispute. It does not have to determine who is right and who is wrong. The purpose of contract law is to enable parties to have the government penalize both if they have a dispute; and contract doctrines merely give parties a reliable way to indicate ex ante their desire for such government involvement, and to limit the size and the variance of the penalty to something close to what should be sufficient: a finger rather than a head.

This theory answers the question, Why, if nonlegal sanctions are so powerful, do people take care to ensure that their contracts comply with the formalities of contract law? The answer is that although nonlegal sanctions are powerful, they cannot deter defections when the benefit from defection is high enough. When this occurs, the potential victim of breach benefits from the contract. And both parties, not knowing in advance whether they will be injured by the price change or benefited by it, agree to the contract in order to protect themselves from defection.

If this view of contract law seems perverse, consider the similarities between the role it assigns to the courts and the historical role of related institutions. An early remedy for legal wrongs was the trial by battle, which entitled the complainant to face the defendant in a tournament.[5] And an extremely important non- or semi-legal institution for dispute resolution—in every major country and in every period of history before the twentieth century—has been the duel. Both forms of dispute resolution present the following puzzle. If the outcome of the dispute depends on skill with arms, and not on the existence of opportunism, then the tournament and the duel does not deter opportunism but gives skillful people a license to do whatever they want, a license that, ironically, they would not want, since then no one would trust them to keep their promises. The history of dueling reveals many practices that took away this advantage, including the practices of giving the challenged person the choice of weapons and, in the case of pistols, forcing the parties to use highly inaccurate pistols at a great distance—or, just as good, accurate pistols at a short distance.[6] The Icelandic holmgang required both parties to wear so much armor that death was rare (Miller 1990). These practices made the outcome of the duel a matter of luck, much as the outcome of a dispute before a radically incompetent court is a matter of luck. So the reason for the persistence of these practices may be that the availability of the threat of joint loss, together with a reputational sanction if the threat is not followed through, can deter the breakdown of a relationship. People do not cheat because they fear being subject to huge losses; they do not settle because they fear being thought to be cowards.

Legal Implications

Form. The value of a legally enforceable promise as a commitment device depends on the freedom of parties to opt into or out of legal liability. The ability to impose legal liability on a stranger without ex ante consent would give people the ability to use courts strategically to extract wealth from each other. This was a problem with the duel: a highly skillful person or a risk-preferring person can threaten to challenge people to duels if they do not do what he wants them to do, and his skill or boldness enables him to overcome the element of randomness. In England, great nobles feared such challenges from lesser nobles, and so had to isolate themselves or surround themselves with bodyguards in order to deprive the lesser nobles of the opportunity to challenge them to a duel. Similarly, if X and three friends can persuasively but fraudulently say before a jury that Y had agreed to buy some stock at time 0 but had since breached his promise because prices had fallen at time 1, then any con artist can use the court system in order to effect wealth-decreasing transfers from victims to himself. To prevent such behavior, contract law must distinguish between obligations that have been voluntarily incurred and fraudulent representations of such obligations.

The mechanism for making this distinction is form. Courts and legislatures establish certain forms, like the seal or the writing, as a way of indicating a desire for legal enforcement; by violating these forms one can indicate one's desire to avoid legal enforcement.

Form, then, takes center stage in the commitment model, whereas in the standard one-shot model it had been elbowed into the wings. Under the standard model it was understood that form is a way for parties to signal their desire to opt into or out of legal enforcement. It was recognized that the cost of forms is that sophisticated people can use them to bind others who believe themselves not legally bound and to avoid binding themselves when others believe them legally bound. The benefit of forms is that they enable people to avoid legal liability.[7] These points remain relevant. But because there was no theory explaining why parties would want to opt out of legal enforcement, there was no explanation for why form mattered.

This is why modern writers on contracts have such trouble with people like Holmes and Hand, who believed that parties' intentions are irrelevant as long as they meet the requirements of forms.[8] The formalist approach seems perverse: why should it matter if parties adhere to a form or not, when we really care about their intentions? If a party fails to dot an i, we shouldn't let that tiny omission prevent us from enforcing the contract. But this modern view assumes away the problem that form is intended to meet—the problem whether or not we know the parties' intentions. It assumes that courts can determine the parties' intentions from context and common sense. If this assumption is correct, then courts should ignore form. But Holmes' and Hand's view make sense under the assumption of judicial incompetence. Courts cannot read parties' intentions from context, so they must rely on the forms that the parties choose. There is no evidence for the modern conviction that judges can determine intentions. And although courts are no longer as formalistic as they used to be, there is no reason to believe that this trend is desirable, that judges are more competent than they used to be, or that contracts are more complex, or that the old attitude was wrong. The modern view is based on an empirical hunch, and no more, and on this basis contract law has slowly shed many of its formal requirements.

The formalist approach requires that courts or legislatures choose the form that parties must satisfy in order to convey their desire that a court intervene if a dispute arises. A historical example is the seal. In order to obtain a legal remedy for breach of certain kinds of promises, the promisee had to produce a document that bore the promisor's stamp in hardened wax. No doubt one could forge seals, but it must have been difficult to do so. Because the promisor would not have placed his seal on the document unless he wanted to make himself vulnerable to legal enforcement, the seal could serve the purpose of form.

The problem with the seal was that it was expensive and cumbersome. Over the years, various substitutes emerged. For certain contracts, a writing and a signature would serve the purpose. Again, one can forge a signature, but doing so is difficult, and laws against forgery increased the risk. For relatively low-value, short-term contracts, the requirements of form are now quite minimal, but they remain significant. The contract must be based on a quid pro quo; there must be evidence of something like an offer and acceptance; the terms of the original contract must be sufficiently clear and definite; and so on.

It is clear, then, that the formal requirements of a contract, or of a kind of contract, can vary across a range, from minimal to maximal. Costly form protects people who do not enter contracts: they are less likely to be obliged to keep a promise they did not make. But costly form also increases the cost of business for those who want to obligate themselves. Cheap form reduces the cost of entering contracts, but increases the risk of being held liable through fraud or by accident. The discussion of gift-giving and the consideration doctrine in Chapter 4 illustrated one such trade-off: if legal enforcement of gift promises does not have much social value, and if a clear quid pro quo is a useful form for identifying intendedly legal promises, then the consideration doctrine makes sense. Of course, the proper tradeoffs will depend on circumstances, will never be obvious, and will change as the underlying incentives change.[9]

Form is self-generating. Commentators frequently urge courts to abandon formalities, and decide cases using standards. Many courts have done this. But efforts to replace formalistic rules with standards fail, because form arises when parties imitate the contracts that courts in the past have found to be legally binding. If a court says that it applies a standard, cautious lawyers will draft the next contract with an eye toward the contract that the court approved, because then in case of legal dispute the lawyers can plausibly argue that since the last contract was enforced, and the current contract is similar to it, the standard that approved the last contract would approve the current contract as well. An example of this is the shift from the common-law mirror image rule to the Uniform Commercial Code's equally formalistic though more complex 2–209, which simply shifts the risk from one party to the other, rather than reducing formality in the way that Llewellyn, the U.C.C.'s drafter, desired (Baird and Weisberg 1982). Other examples include the Statute of Frauds, the parol evidence rule, and the consideration doctrine, which persist despite the assaults of judges and commentators. When judges relax the requirement of form in a particular case, usually to ensure liability when one party appears to have reasonably believed that a contract had been created, the new fact pattern becomes itself a form to which future parties adhere.

In a way, then, judicial mistakes are self-correcting. If a court finds a person contractually liable who did not actually observe the formal requirements, then next time the person will simply take steps to be clearer. He will issue disclaimers every step of the way, as occurred in the wake of *Hoffman v. Red Owl.*[10] This becomes the new formality. If, however, courts make the formal requirements too costly, then people will not be able to opt into judicial enforcement. The instability of formal requirements over time has been noted by many commentators (Kennedy 1973, Johnston 1991, Hadfield 1992).

Damages, excuse. Suppose that Seller and Buyer enter a contract in which the payoff for (S, B) from mutual cooperation is (100, 100), the payoff when Seller defects and Buyer cooperates is (110, 50), and the payoff when both defect is (70, 70). If the court had perfect information, it might award damages equal to 50 against the person who defects when the other cooperates, as this would compensate the loser while deterring future cheaters.

Suppose that Buyer sues for failing to deliver, and claims damages of $X. Seller countersues, claiming that Buyer breached by failing to provide specifications, and claims damages of $X. Suppose further that Seller and Buyer believe that the court will randomly choose an amount in this range, that is from -$X to $X; and that each can improve his chance of winning $1 by spending $1 on litigation. Then Seller and Buyer will each spend $2X in litigation in order to avoid a loss of $2X. If, for example, Seller sues for $50, and Buyer countersues for $50, they would each spend $100 in litigation costs.

Observe that the $100 expected cost for each party, C, serves the purpose of commitment. If each party can credibly impose a $100 litigation cost on the other if cheating occurs, the ex ante payoff from defection is $10 or -$30, which is lower than the ex ante payoff from cooperation when the other person cooperates, which remains $100. Ex post, the parties face a strong incentive to settle for an award of $0, rather than spend $100 each for an expected gain of $0, but they resist this incentive in order to maintain a reputation for toughness.

Although this result is good, it is not optimal. Commitment would be possible if the parties could agree not to spend more than $11 to litigate, since this amount exceeds the $10 payoff from defection. When parties are unable to settle, they will spend $200 when they should spend no more (or less) than $20, and ex ante an agreement that required a joint payment of $20 would maximize value. But the parties could not make such an agreement ex ante; and ex post any agreement to settle might be taken by observers to indicate that the parties are not tough, and can be exploited in the future. This would be like two duelists agreeing that they will fire at each other with cap guns.

Several features of contract law mitigate this problem. Contract doctrines strictly limit the size of damage awards (Schwartz 1990). The refusal to award damages when the loss is speculative or a result of emotional distress is an instance of this, as are the limitation on consequential damages and the penalty doctrine. The ordinary rules of damages are almost always undercompensatory. Thus, the parties would not expect the court to award them between -$50 and $50, but to award them an amount in a narrower range. Thus, they would not spend $100 each on litigation, but say $90, $80, or less.

In addition, the excuse doctrines, which release the promisor from his obligation under certain conditions, limit contract liability, and these doctrines serve this desirable purpose even if, consistent with my assumption of judicial incompetence (and, many would say, with the case law), they are applied randomly. The reason is that if the range is, say, -$80 to $80, the excuse doctrine means that occasionally courts will award $0 rather than -$80 or $80. Thus, the average award declines, and so will C. The fact that the restitution will often be awarded reduces the variance that would otherwise result and be painfully felt by risk-averse parties.

All of this might seem like awfully rough justice, but that may be the best we can do in an imperfect world. Dueling was also rough justice. The rules of dueling, including the use of seconds to intervene, the insertion of elements of chance to even out differences in skill, and the reliance on weapons that reduce the risk and amount of harm, are elements that, like contract doctrine, preserve the deterrent effect of the institution while ensuring that harm that occurs when deterrence fails does not exceed by too much the amount necessary for the deterrent effect. Contract law is best understood as a modern version of dueling. It is an incremental advance on dueling that preserves the essential character of that institution while eliminating violence and physical harm, and reducing the variance of outcomes.

The treatment of custom. If one asked merchants why they put particular terms in their contracts, they would sometimes say, "because they are important to me," but they would usually say, "because it is the custom." Most contractual terms, and contractual behavior, are a matter of custom, including whether the seller delivers or the buyer picks up the goods, the allocation of the risk of loss when the goods are in transit, the form and negotiability of the instrument used to transfer funds, the appropriateness of breach or renegotiation, and sometimes even price. Scholars waver between praising custom as the reflection of accumulating commercial wisdom, and condemning it as a drag on commercial advance.[11]

What is custom? This term is used in a variety of ways, and I will pick out just two of them. People speak of customs arising between two parties, and customs arising in groups. Let us start with the former.

Suppose that a buyer and a seller have a long-term contract according to which every month the seller supplies the requirements of the buyer. Imagine that the buyer informs the seller of its requirements on the first of each month, and that, although the contract says vaguely that the seller must deliver within a "reasonable time," over several years the seller always delivers on the 10th of each month. After a while, the buyer starts to rely on this regularity, and modifies its business practices—for example, by failing to have alternative suppliers lined up if the seller is late. Then one day the seller delivers after the 10th, the buyer suffers losses, and it sues the seller. The question is whether the seller's practice of delivering on the 10th ripened into a "custom," or what the Uniform Commercial Code would call a "course of dealing."

The buyer might argue that the delivery on the 10th became a term of the contract, while the seller would argue that it delivered on the 10th just because that was a convenient time for it. How would one resolve this dispute? Some commentators argue that the court can discover whether a custom exists, perhaps by interviewing experts or observing behavior in the industry (Cooter 1994b). But in our example, the only relevant players are the buyer and the seller. Each had a different understanding, and none of the facts tells us whether a custom exists or not.

Now, if judicial enforcement were costless and perfect, or if the parties could rely on cheap and accurate merchant courts, they could resolve this problem just as one resolves any contractual dispute under the traditional model. If delivery on the 10th maximizes the ex ante value of the contract, then the seller should be required to deliver on the 10th. Whether it does simply depends on the relative cost to the seller and benefit to the buyer. If, for example, the seller can cheaply ensure that delivery always occurs on the 10th and the buyer gains a great deal by this commitment, then the cheap and accurate court should hold in favor of the buyer. The brute fact of the behavioral regularity does not tell us whether the regularity reflects value-maximizing conduct or accident. The idea of "custom" does no work beyond the traditional analysis, and if real courts are too incompetent to determine optimal terms in one-shot contracts, then there is little reason to believe that they can determine the optimal terms in a long-term contract.

People say that customs spread among groups of transactors. They seem to assume that because custom transcend the practices of any pair of individuals, they are more recognizable by courts and hence more appropriately enforced. How might this work? When different pairs hit upon similar customs, and these pairs do better than others, one might expect partners with bad customs to abandon them and to mimic the customs of the people who are doing well. Thus, valuable customs spread. If people simultaneously deal with many others, or switch partners a lot, one would expect customs to spread more rapidly. Call this group of people from which an individual takes his partners a "trade group." Good customs might spread within the trade group, because people with worse customs would imitate those with better customs. However, it is quite possible that suboptimal customs will prevail, especially if the initial experiments are poor ones, and a critical mass of people with bad customs is reached before anyone hits upon better customs. If the trade group has competition from other trade groups, then its members may be driven out of business or they may experiment with new customs; but this all depends on the level of competition, if any, and the extent to which parties can distinguish practices by the amount that they contribute to the total value of the relationship.

PART THREE

Normative Implications

10

Efficiency and Distributive Justice

Although earlier chapters focused on descriptive claims about the relationship between law and social norms, normative judgments tumbled in now and then. This chapter and the next two chapters gather them together. The goals of these chapters are modest: not to convince you of the value of any particular legal reform so much as to convince you that common normative judgments in legal analysis should take account of complexities of nonlegal regulation more often than they do. This chapter focuses on the efficiency of group and extra-group norms, and the relationship between norms, welfare systems, and distributive justice.

Moon Cakes

An article in the Wall Street Journal (C. Smith 1998) describes a ritual that, despite its foreign provenance, is intensely familiar. It concerns the ritual exchange of the moon cake.

> Years ago the quarter-pound cakes, most often filled with a paste of mashed lotus root, sugar and oil, were prized gifts, a rare rib-sticking treat that would keep well into the icy winter months when most people subsisted on cabbage. But the urban Chinese are richer now, and moon cakes have become more bother than bounty. Like Christmas fruitcakes in the U.S., they get passed from person to person until the festival ends—and the last one holding the cakes has to eat them or quietly throw them away.

The moon cake had once been a meaningful gift. When people dined on cabbage every day, a moon cake must have provided some welcome variety. The puzzle is why people gave away their moon cakes, in return for more moon cakes, rather than consuming the moon cakes that they had bought or produced themselves. Transaction costs alone would have made such exchanges inefficient (a problem solved currently by a thriving secondary market of moon cake coupons). The answer to the puzzle is that the gift of a moon cake was a signal that people conveyed to their friends, relatives, and associates, to show that they were good cooperators. Like other non-monetary gifts, the moon cake was both costly to the donor and not so valuable to the donee that it would swamp the cooperative benefits of a relationship.

> Moon cakes from famous shops such as Xinghua Lou get passed around the most. Ms. Li believes many boxes pass through five or six hands ahead of the "best before" date stamped on the side. One recent newspaper article recounted the case of a man who was given the same box of moon cakes that he had given away weeks before.

Why is the signal a gift of moon cakes, rather than some other item? The answer is that people give each other moon cakes this year because people gave each other moon cakes last year. At any time *t*, people must act in conformity with expectations based on time *t*-1. If they do not, then other people would begin to doubt whether they intend to continue a relationship, just as Americans wonder whether the failure to send a Christmas or holiday card this year, after sending one last year, is meant to be a signal.

The origin of the moon cake exchange is shrouded in the mists of time—the tradition is a thousand years old—but part of the explanation for the ritual is surely that the holiday during which moon cakes are conveyed takes place on the full moon closest to the autumnal equinox. And the holiday celebrating the harvest occurred during the full moon presumably because that day could be readily identified year after year by farmers who witnessed the location of the sun and the phase of the moon but did not have written calendars. People could have given each other pieces of clothing, or other kinds of food, but the moon cake was suggested by the holiday, which itself was suggested by the salience of a moment during the harvest. The moon, and hence the moon cake, was a focal point around which people coordinated their signaling behavior.

> Wang Yafang opens her first box of the season and cuts a guest a quarter cake . . . The 53-year-old housewife expects to receive more than a dozen boxes before the festival is over, though her family manages to chew through only one box each year.

When prosperity came, the moon cake lost its appeal. People did not want moon cakes, because they could purchase more delicious pastries at their local bakeries and anyway they now had a more varied diet than in the past when cabbage dominated their meals. Still, no one could deviate from the thousand-year-old equilibrium. Given that people expected moon cakes from their friends and associates, disappointment of this expectation could only be interpreted as an indication that the relationship was at an end.[1] To avoid this inference, people continued to give moon cakes to people who did not want them, and received in return moon cakes that they did not want either.

> "I'll send two boxes to my little brother's family, two to each of my husband's brothers, two to my parents . . ." she says, counting down how she will dispose of them all. "Somehow, we always end up with one box that we can't give away," she says with a sigh.

Although no one likes the taste of moon cakes, a tremendous quantity of them is manufactured (21 million by just one of many bakeries that make them) and vast sums are spent on them. The ritual may finally end, yet the conclusion is inescapable that it has persisted long beyond the point at which it ceased having any value, if it ever did have any value. The comic pathos of the situation is the reverse of the despair produced by dowry competitions in India, the difference being that trends in technology and demographics have made the signals too cheap rather than too expensive.

Social Norms and Efficiency

One finds in the literature two kinds of claims about the efficiency of social norms. First, many economists and law professors assume that social norms solve strategic dilemmas that would otherwise reduce overall well-being. An example is Frank and Cook's (1997) argument that people overconsume positional goods in a competition that leaves everyone or almost everyone worse off than he would be if the competition could be restricted. Frank and Cook suggest that disapproval of cosmetic surgery, conspicuous consumption, and similar behavior reflects welfare-enhancing social norms.

There are a number of difficulties with this claim. First, one must distinguish between attitudes and behavior. An individual might envy or feel contempt for wealthy or beautiful people, but it does not follow that the accumulation of wealth or the purchase of cosmetic surgery violates a social norm. More to the point, even members of the elite may recognize that the competition over beauty and wealth may make everyone worse off, but if no one can break out of the competition, then it serves no purpose to say that the behavior violates a social norm. It is thus useful to limit the use of the concept of social norm to cases where people punish those who engage in the proscribed behavior—for example, by avoiding them—and there is little evidence that people in the United States try to sanction wealthy and attractive people in this way. Otherwise, one could not understand how people could condemn the moon cake exchange ritual while engaging in it. Yet people often feel trapped by social norms that they abhor.

Second, one must be careful about *which* people hold the attitudes under discussion. Certainly, there are social groups in which conspicuous consumption or the purchase of cosmetic surgery would provoke sanctions. A member of an old-order Amish community will be ostracized if he or she violates prohibitions against the display of ornaments. But I doubt that most Americans would take steps to sanction individuals who engage in conspicuous consumption or undergo cosmetic surgery.

Third, there remains a matter of interpretation. Suppose that Americans sanctioned individuals who undergo cosmetic surgery. They might sanction these people simply because they envy them and not because they realize that anyone who undergoes cosmetic surgery promotes a self-defeating competition for beauty that reduces welfare in the aggregate because beauty is a positional good. The former interpretation is more plausible. On the other side, one can imagine circumstances under which social norms against cosmetic surgery do arise, even if cosmetic surgery is not a positional good, and instead gives pleasure both to the patient and to most other people. I will say more on this possibility below.

Functionalism—the view that social practices and norms are efficient or adaptive in some way—is empirically false and methodologically sterile. To answer the question of whether a social norm is efficient, one must produce a theory about the supply of social norms as well as the demand. There are two main sources of supply, each corresponding to the two main kinds of norms—group norms and extra-group norms. I discuss each kind below.

Efficiency of Group Norms

Some scholars, notably Ellickson (1991), argue that social norms are efficient when they arise in close-knit groups. Ellickson advances this argument mainly as an empirical claim. Ellickson's study of cattle ranchers and farmers in Shasta County, California shows that these people do not rely heavily on the law and do not even have a good idea of what the law is, yet they cooperate in impressive ways. Neighboring landowners repair fences, retrieve stray cattle, pay for damage done by their cattle to the property of others, keep their promises, and pay their debts. These findings resemble the findings of studies in the literature on common pools (Ostrom 1990). This literature shows that individuals in a remarkable array of settings—Turkish fisheries, Alpine pastures, Japanese farming regions—cooperate in the production of collective goods in the absence of an effective legal regime. These studies decisively contradict the view that people will never cooperate unless so required by law.

As an aside, I should observe that the studies do not show that the observed cooperation is optimal. True, the farmers fix the fences rather than letting them fall to pieces, but optimal cooperation might require more than that. The standard model of cooperation in teams shows that if each party's individual interest in the collective good is strong enough, the parties will produce some of it, and cooperation will fail just on the margin. The parties repair fences but not as quickly as would a single owner of the fences. They return a neighbor's stray cattle but not as rapidly as they would recover their own. Similar criticisms can be made of the studies of common pools (for details, see E. Posner 1996a).

Let us turn to the theoretical reasons for believing that group norms are likely to be efficient or inefficient. Ellickson does not make the functionalist mistake of assuming that because cattle ranchers want efficient norms, they will produce such norms. To explain the supply of efficient norms, he appeals to the theory of repeated games. This theory shows that two people in a repeated prisoner's dilemma might engage in the optimal level of cooperation. But as Ellickson acknowledges, the theory does not show that optimal cooperation is necessary. A farmer who shares a length of fence with another farmer might rationally or mistakenly make repairs too rarely; if the other farmer retaliates by acting in the same way, a suboptimal equilibrium will result. Still, optimal cooperation between two people seems likely.

Extension of the two-person game to the n-person game, however, is fraught with difficulties. Suppose a rancher's cattle stray onto the land of several neighbors who themselves do not own cattle. There is an optimal level of cooperation, but the form of cooperation may be complex and difficult to realize. Suppose that the optimal level of cooperation is achieved when the rancher expends x units of effort to restrain his cattle, each neighbor expends y units of effort to return his cattle, and the rancher compensates the neighbors either by making occasional cash payments or repairing fences or helping out in other ways. True, a neighbor might meet his obligations and expend y units of effort, but he might instead let the rancher's cattle wander onto someone else's land. The rancher's threat of retaliation might not be sufficient to enlist his cooperation, and the other neighbors might not be able to coordinate to punish the shirker. Everyone has an incentive to free-ride, resulting in a low level of cooperation or none at all. This is, of course, the collective action problem. Although game theorists have shown that in principle certain equilibrium strategies could result in n-person cooperation, these strategies seem implausible (see Chapter 2.)

Let me now turn to the signaling model. To understand the efficiency implications of the signaling model, begin by imagining that information costs are zero. Because everyone knows the type of everyone else, and because signaling is costly, no one sends signals. Good types match up with other good types, and bad types match up with other bad types or not at all. After players match up, some amount of cooperation occurs within those matches. Not even the assumption of perfect information guarantees that the optimal amount of cooperation would occur, but let us assume that such a result is plausible. In the repeated prisoner's dilemma without information costs, each player knows that if both players choose a sufficiently cooperative strategy, like tit-for-tat, both will do better than if both choose a less cooperative strategy, such as cheat-but-then-play-tit-for-tat. The cooperative surplus—which might consist of money profits (Chapter 9), the pleasures of friendship (Chapter 4), the generation of political influence (Chapter 7), the marital surplus (Chapter 5)—is the same thing as the internal collective good, a concept introduced in Chapter 2.

Information costs are not zero. When information costs cross a threshold, good types signal in order to distinguish themselves from bad types. Depending on the relevant parameters, signaling may or may not produce separating equilibria, but the relevant point is that there is no reason to believe that any given equilibrium is likely to be more efficient than an alternative. When people signal, they impose costs on third parties, so they do not have the proper incentives to engage in efficient signaling.

To see why, imagine a pooling equilibrium in which no one sends a signal but in which "receivers" (those who receive a signal) cooperate with everyone. Receivers would cooperate as long as the expected gains from cooperating with the good types minus the expected losses from cooperating with the bad types exceeded the value of their alternative opportunities. One might observe in such an equilibrium some or a fair amount of cooperation but not necessarily optimal cooperation. Now suppose that exogenous changes supply the good types with a technology that enables them to distinguish themselves from the bad types by issuing differentially costly signals. If the good types start signaling, the receivers might stop cooperating with the bad types and cooperate solely with the good types. Because the receivers would devote more time to their relations with good types, it is possible that the gains from cooperation would increase for the good types, justifying the good types' investment in the signals.

When the good types engage in signaling, then, they increase the value of the internal collective good obtained by themselves and the receivers but they also decrease the value of the internal collective good obtained by bad types. If this latter decrease is small and the number of bad types is low, then there will be an efficiency gain. Otherwise, there will be an efficiency loss. And if the bad types are able to mimic the signal, and do so in order to avoid being identified as bad types, in equilibrium signaling costs are incurred without any offsetting informational gains. Because of these problems, we cannot say at this level of abstraction whether the separating equilibrium produces greater social wealth than the pooling equilibrium does. The value of the internal collective goods in the new equilibrium may be greater or less than the amount produced in the old equilibrium.

The emergence of signaling has a similarly indeterminate effect on what was called the *external collective good.* The signaling technology, whatever it happens to be, might cause the good types to signal their type by aiding others or hurting them. If people signal their type by voting, giving philanthropic gifts, complying with underenforced laws, or volunteering in soup kitchens, then it is possible that the separating equilibrium will have desirable attributes. Whether in fact we like the behavior that emerges in that equilibrium depends on what people would otherwise do and the exact nature of the signal. If they vote without informing themselves or substitute from support of worthwhile charities to support of less worthwhile but more visible charities, then the signaling behavior will be objectionable. If people signal their type by shunning minorities or by engaging in self-censorship, then the separating equilibrium is again inferior to the initial pooling equilibrium. If people signal to each other by exchanging moon cakes, and everyone participates in this activity, then no information is revealed while resources are squandered. If farmers signal by ostentatiously repairing fences, then too much fence-repairing will occur.

When everyone signals, so that a pooling equilibrium exists, there is no revelation of information and thus no contribution to an internal collective good. If, at the same time, the signal does not contribute to an external collective good, then the equilibrium is unambiguously bad, because the cost of signaling is incurred without the production of offsetting benefits. And if the signal actually injures third parties, as with the case of racial discrimination, then the pooling equilibrium is also bad. As noted above, however, other cases are ambiguous. When everyone signals but the signaling contributes to an external collective good, the equilibrium is desirable, as long as the value of the external collective good exceeds the cost of the signaling. When a separating equilibrium results, information is conveyed, and internal collective goods are usually created (unless good types signal too much in order to distinguish themselves from bad types). But the external collective good may be desirable or not, depending on what precisely it is.[2] The following table lays out the analysis:

	Signal's meaning	Internal collective good	External collective good/bad
No one signals	None	Maybe	No
Separating	Meaningful	More likely for good types than for bad types	Yes
Everyone signals	Empty	Maybe	Yes, with more produced than under alternatives

If the table seems insubstantial, it demonstrates why one cannot say much in the abstract, and cannot generate insight without applying the framework to particular social practices.

Even if people do not like a particular equilibrium, it is never clear that legal intervention will improve the situation. If the law taxes or punishes people who issue the signal that supports a particular separating equilibrium, people might substitute to a better or worse signal, or stop signaling and pool. Any of the resulting alternative equilibriums might be better or worse than the initial equilibrium. Frank, McAdams, and others who advocate the use of taxes to deter overconsumption of positional goods overlook the possibility that people will simply substitute to other forms of behavior, equally destructive but harder to tax—for example, weight lifting rather than cosmetic surgery, conspicuous leisure rather than conspicuous consumption.[3] Moreover, because of the self-reinforcing nature of beliefs in signaling equilibriums, incremental changes in the law may fail to change behavior or may cause massive and unpredictable cascades, frustrating efforts to use the law to fine-tune people's actions. This was a theme of Chapters 5, 6, and 7, which discussed examples of this phenomenon.

In sum, one can make no presumptions about whether group norms are efficient[4] and about whether a legal intervention will improve behavior in close-knit groups. Whether it will improve behavior depends on the equilibrium that exists within those groups, which may be more or less efficient, and on the competence of legal institutions that intervene. Ellickson's cattle ranchers engaged in a mixture of cooperative and signaling behavior. Legal intervention may have made things worse, but we really cannot tell. This is not a terribly exciting conclusion, but is, I think, accurate. At lower levels of abstraction, where some institutional detail is filled in, more concrete conclusions are possible, as I tried to show in prior chapters. Given a detailed knowledge of a piece of life, we may be able to predict that the law will improve behavior, because we believe that the participants are not obtaining the cooperative surplus that is available, or are squandering resources in signaling games. When people engage in actions that seem either trivial (flag protection, moon cake exchange) or disproportionate (dowry competitions, ethnic violence, witch hunts) to the resources at stake, we have a clue that dysfunctional signaling prevails and that legal intervention may be desirable. Unfortunately, it is at these times that right-thinking government officials are least able to resist the tide of opinion. It is only when finally everyone believes that the practices are ridiculous that the social norm finally dissolves.[5]

One form of signaling that has always been with us, and has always aroused strong passions, is that of conspicuous consumption. Hirschman (1982, pp. 47–8) notes that even Adam Smith criticized people for buying "trinkets of frivolous utility." Smith also argued that feudal lords acted foolishly by giving up their feudal rights in return for money with which to purchase "trinkets and baubles, fitter to be the play-things of children than the serious pursuits of men." The moralistic attack against unnecessary consumption has been launched by countless philosophers, theologians, social critics, historians, and (like Smith) economists who otherwise are reluctant to criticize personal preferences and who normally celebrate the expansion of markets. The signaling game makes clear that even taking preferences as given, people would be better off if they could exit signaling games without suffering irreparable harm to their reputations. Indeed, powerful ascetic traditions in every major religion show that people understand that they might be better off without any material goods than in a game that compels frenzied accumulation. It is not that people's preferences are wrong; it is that they are hard to satisfy in a society where information asymmetries compel investment in signaling.

Efficiency of Extra-Group Norms

Chapter 3 briefly mentioned evolutionary game theory, which has been used to explain the emergence of norms that govern interactions between people who may be complete strangers, with no information about each other and no expectation of future interaction. Recall the examples from Chapter 3. People drive orthogonally toward intersections at high speeds and receive low payoffs if they crash into each other, and high payoffs if they manage to avoid each other. People could adopt a variety of strategies; for example, "yield to a car coming from the right," "yield to a car coming from the left," "yield to a larger car," "yield to a faster car," and so on. Because no one has any reason to adopt one strategy rather than another, assume that people adopt strategies at random. What would happen?

The first point to observe is that the game is a version of the simple two-person coordination game. If a person believes that all others "yield to a car coming from the right," then this person will adopt the identical strategy and never deviate. If he deviated, he would crash into another car.

The next question is how everyone could arrive at the same belief. For simplicity, imagine that people choose only between two possible strategies, "yield to a car coming from the right" (which I will call Right) and "yield to a car coming from the left" (which I will call Left). Suppose that, by chance, initially 60 percent of the drivers choose Right and 40 percent choose Left, and that everyone maintains whatever strategy he starts with unless he crashes into someone else, in which case he switches to the other strategy. Players start out by randomly matching with each other. Because of the asymmetric starting conditions, people who play Left will crash and switch strategies more often than people who play Right. Thus, over time the number of Lefts will decline and the number of Rights will increase. Eventually, everyone will play Right, and the norm will be in place.[6]

As the example shows, the outcome depends on the focal character of the strategy and on initial conditions. Right and Left are focal in the sense that if everyone follows the same strategy, it is always clear what to do, and the desirable outcome will be achieved. In contrast, the strategy of yielding to the faster driver will produce ambiguous situations whenever the drivers cannot tell who is driving faster. That the initial conditions matter can be seen from the fact that if people had begun by choosing Left more often than Right, then Left would have become established as the norm. But because focal points depend on accidents of history and psychology, and the initial conditions are random, one cannot assume that equilibrium strategies will be efficient. The norm of yielding to the faster car could be more efficient than Right or Left once the problem of ambiguity is solved. If state intervention could solve the ambiguity, then it would improve welfare beyond the level achieved through evolution.

A clear example of the way state intervention has raised welfare beyond the level produced by the evolution of norms comes from another coordination game, this one involving conflict between two strangers over a bundle of resources. When two strangers come upon something of value, they can play Aggressive or Passive. Aggressive means seizing the goods, either without injury (if the other player plays Passive) or with great injury and only possible success (if the other player plays Aggressive). If both players play Passive, time is wasted and the goods deteriorate. In the absence of a norm, the optimal strategy is to randomize between Passive and Aggressive. Suppose, however, there exists a possession norm that states that the person with possession plays the aggressive move, the person without possession the passive move. Such a norm is stable in the sense that once the beliefs that support it exist, no one has an incentive to deviate from it. An analysis similar to the analysis applied to the driving game shows that people randomly choosing the possession norm would do better than people who randomly chose the mixed strategy, and over time the possession norm would prevail, just as Right did in our example above. (See Sugden 1986, Hirshleifer 1982, Young 1998a.)

The possession norm yields gains over anarchy because it allows the parties to avoid the bad outcomes of mutual passiveness and mutual aggression. However, state intervention improves welfare further. The problem with the possession norm is that it does not protect property in which a person has invested resources but which he cannot overtly possess—for example, large areas of land or chattels that are best used by third parties. The solution to this problem is a recording system, such as those used for real estate transactions and security interests in personal property. But a recording system cannot *evolve* in the way that the possession norm has evolved. It must be self-consciously created by an agency, such as a legislature.[7]

This is only a sketch of a complex analysis, but further detail is not necessary for the purpose of drawing an important general lesson. Extra-group norms, like group norms, should be understood as descriptions of equilibrium behavior in games in which people with private information interact with each other. Casual analysis of these games shows that social value will not necessarily be maximized and that state intervention can potentially improve payoffs for the players. Of course, whether state intervention really will make people better off, rather than worse off, depends heavily on circumstances.

Stigma and the Redistribution of Wealth

What is the best way of transferring wealth to the poor? Most economists seem to agree that a desirable system would have the following three characteristics. First, the system would be based on transfers funded from tax revenues rather than alternatives such as minimum wage laws and legal rights biased in favor of the poor (Kaplow and Shavell 1994). Minimum wage laws prevent employers from hiring people with skills that generate revenues insufficient to cover the mandated wage. If these laws were abolished, employers could hire low-skilled people, and the wealth that would be generated from the employment could be taxed for the purpose of aiding the poor. Minimum wage laws and similar laws perversely transfer wealth from the very poor to workers at the margin and produce a deadweight loss to boot.

Second, the welfare system would distribute cash rather than in-kind benefits. The reason is that recipients prefer a fungible item like cash to goods, because cash can be more easily traded for whatever the recipients value the most. By now it is familiar that this argument assumes that the purpose of welfare law is to enhance the utility of the poor, when in fact welfare serves more complex purposes, such as controlling consumption. These purposes may justify the use of in-kind benefits, such as food stamps, medical care (in the form of mandated emergency room assistance), housing assistance, utility subsidies, disability insurance, and pensions. Still, this explanation has an air of rationalization. No one has explained why the current mixture of cash and in-kind benefits is superior to alternatives, and the modern welfare system's frequent use of in-kind transfers remains puzzling.

Third, the welfare system would distribute benefits in a way that would minimize perverse incentives not to work, not to invest in education, to take risks, to have children whom one cannot support. Historically, there have been two approaches to these problems. The first approach is to make the benefits so meager that most people prefer working and earning money to not working and receiving welfare. The potential recipient is required to bear some of the cost of not working, just as insurers use deductibles to force clients to bear some of the risk of loss of insured goods. The second approach is to limit benefits to those whose poverty results from misfortune rather than from opportunism: more technically, poverty caused by a contingency that one cannot prevent or insure against except by paying a premium so high that it would throw one into poverty. According to this approach, benefits (or more generous benefits) go to those, like the physically disabled, who can prove that they cannot work, and to those who belong to classes, like children and unmarried mothers, who are generally not in a position to work.

These three issues of welfare policy—transfers as opposed to rules biased in favor of the poor, cash benefits as opposed to in-kind benefits, and the incentive problem—are all connected to the phenomenon of *welfare stigma.* Welfare stigma occurs when people shun or ostracize those who receive welfare transfers from the government. The phenomenon is well-documented, and has been a matter of concern to sociologists, social workers, and policy makers for decades (for example, Waxman 1977).

To understand why people stigmatize welfare recipients, we can use the signaling model described in Chapter 2. Poor people can be divided into good types and bad types. The bad types are poor mainly because they do not invest in education or training or insurance, do not save money, and do not cultivate relationships, instead preferring short-term gains over the long-term benefits that accrue to those who exercise prudence and self-restraint. The good types are poor mainly because of bad luck: they are the victims of uninsurable risks such as economic dislocation, illness, natural disaster, and crime. Suppose that in a relatively prosperous and stable economy the fraction of the poor belonging to the bad type exceeds the fraction of the non-poor belonging to the bad type. If this is so and poverty is observable, observers rationally infer that a poor person is more likely to belong to the bad type than a non-poor person is. Employers refuse to hire the poor person, shop owners refuse to sell to him, society turns its back on him.

Poverty is not always observable. The poor good types have powerful incentives to avoid being identified with the poor bad types, and will take steps to avoid this result. Mainly, they will mimic the behavior of the non-poor. They devote resources to dressing respectably, imitating the manners of the non-poor, maintaining a neat dwelling, and trying to move away from the poorest areas of town. If they can separate themselves observably from the bad types, they enhance their opportunities for valuable cooperative relationships with the non-poor.

Even if the good types fail to conceal their poverty from observers, their efforts to signal can, by themselves, cause observers to infer that they are good types. Indeed, people have long divided the poor into the "respectable" or "deserving" poor, and the undeserving poor. The respectable poor incur costs to show that they belong to the good type, and observers are more willing to interact with them socially and commercially than they are with the others, whom they cannot trust. But signals are often blurry. Observers are likely to stigmatize obviously poor people; and will cooperate only with those poor people who successfully conceal their poverty or send persuasive signals that they belong to the good type.

So poor people send signals in order to show that they belong to the good type, and those who fail to send these signals are shunned or stigmatized by respectable society. Compounding this problem is that people at the margin between poor and middle class will avoid associating with the poor in order to demonstrate that they are closer to middle class than to poverty. People shun the "undeserving" poor because they do not trust them, and they shun the deserving and undeserving alike because they (the non-poor) want others to believe that they belong to the good type.

Welfare programs can enhance or undermine welfare stigma, depending on their design. Consider poor people who successfully conceal their poverty. If the government gives them food stamps, rather than cash, the recipients cannot buy food without revealing their poverty. Certain poor people who would otherwise be able to conceal their poverty must reveal themselves, and risk being stigmatized. Similarly, when welfare programs require recipients to visit public offices, allow social workers into their homes, and engage in other identifying behaviors, they stigmatize the recipients. Stigmatizing welfare programs force the recipient to liquidate his reputational capital. When the refusal to accept charity is a signal that one belongs to the good type, publication of a person's receipt of charity reveals that he belongs to the bad type. The recipient's loss of reputation results in a loss of future income (employers are less likely to hire a former welfare recipient), and in future consumption of non-market goods (people are less likely to socialize with a former welfare recipient). In return, he receives short-term benefits from the government, and he becomes dependent on further handouts precisely because every handout decreases his attractiveness to potential employers.

Now let us return to the three elements of welfare policy that were discussed above, addressing them in reverse order. What is the connection between the incentive problem and stigma? The answer is that historically governments have exploited welfare stigma in the hope that it would diminish the incentive problem. The theory behind the workhouse system in England was that by supplying food and clothing, the workhouse system prevented extreme poverty and misery, but by publicizing the identity of recipients (and by controlling their behavior) it discouraged people from applying. To receive benefits, most people had to leave their community and enter workhouses, where their movements were restricted and where they were required to wear uniforms. "Until the last decade of the nineteenth century the central authority consciously fostered the idea that the workhouse carried social disgrace, and middle-class observers naturally accepted this. Guardians themselves tried to emphasize the disgrace of pauperism: some of them regularly published lists of names to be pinned to the church door and other conspicuous places." (Crowther 1981, p. 255.)

The problem with this system was that because entry into the workhouse was *so* stigmatizing, the "deserving poor" refused to enter the workhouse, except when truly desperate, ill, or old, whereas the "casual" or undeserving poor—for whom no one had any sympathy and who had no reputation to lose—were the major beneficiaries. According to one contemporary observer, "The aversion to the 'House' is absolutely universal, and almost any amount of suffering and privation will be endured by people rather than go into it. Loss of liberty is the most general reason assigned to this aversion, but the dislike of decent people to be compelled to mix with those whose past life and present habit are the reverse of respectable is also strongly felt." (Crowther 1981, p. 240, and see generally chs. 8–9.) In the 1920s, when the ranks of the poor received the veterans of World War I, people were forced to reconcile their prior beliefs that poor people were bad types with their new beliefs that the veterans of the trenches were good types—indeed, patriots and heroes. The first set of beliefs yielded to the second. The modern welfare state in England was motivated, in part, by a desire to eliminate the stigma of poor relief. Similarly, the modern welfare state in the United States arose during the Great

Depression, when people's prior beliefs that poor people were bad types gave way to their beliefs that middle class people, newly impoverished, belonged to the good type.

The decline in the reliance on stigma in welfare policy probably resulted from several factors. First, stigma deterred the more sympathetic people from receiving welfare or humiliated them, without affecting the decisions of less sympathetic people. Second, stigma created a class of outcasts, who would have trouble obtaining future employment. This problem parallels the danger of using public sanctions to stigmatize people who have committed crimes (see Chapter 6). Third, the degree of stigma depends on diverse factors that are hard to identify and harder to control, such as the number of people out of work and the beliefs of the general population about the kind of person who does not work. A welfare system that deters opportunism by relying on stigma may therefore produce unpredictable and perverse incentives.

This argument raises the question of how welfare can be distributed in a non-stigmatizing way, and allows us to return to the connection between the *form* of welfare transfers and stigma. As noted above, some welfare transfers identify recipients and others do not. The workhouse system identifies recipients by removing them from the community and requiring them to wear uniforms. Modern welfare systems identify recipients by requiring them (sometimes) to allow social workers into their homes or to visit public offices where they may be seen by neighbors. The clearest example of an existing stigmatizing welfare benefit is the food stamp program, which has generated a folklore of welfare recipients using food stamps to buy junk food.

In contrast, the social security system does not reveal who is poor and who is not. Because everyone must participate in the system, the knowledge that someone receives social security checks does not allow one to infer anything about that person's wealth. Subsidies for electricity and other utilities also do not identify recipients to most outsiders. Consider also a hypothetical food stamp system, under which everyone received food stamps from the government, presumably after a tax hike equal to the average food expenditure, or under which non-poor people could buy food only with food stamps which they could purchase from government offices with cash. Because under such systems everyone would use food stamps, the use of food stamps would no longer distinguish the poor and the non-poor. Finally, housing assistance programs are stigmatizing when they require people to live in designated low-income buildings and they are not stigmatizing when they merely subsidize rental payments.

Many puzzling forms of welfare are best understood as methods of aiding the poor without stigmatizing them. Minimum wage laws conceal the identity of their beneficiaries. The minimum wage law condemns to poverty and unemployment people who have few skills, but these people would receive welfare even if the minimum wage law did not exist. By increasing the income of the marginal workers, the minimum wage law makes it unnecessary to distribute welfare payments to them and thus spares them the stigma of the welfare recipient. This means that poor good types who would refuse welfare benefits that identify recipients actually receive disguised transfers in the form of inflated wages. Rent control laws also conceal the poverty of their beneficiaries. Because rich and poor people alike benefit from rent control laws (unless they are means-related), observers cannot infer a person's wealth from his residence in a rent-controlled apartment. Thus, while one must acknowledge the harms caused by these and similar laws, they also have a benefit: they reduce the power of outcast status by obfuscating the boundary between poor and non-poor. They do increase the stigma incurred by some people—the people who are too poor to benefit from rent control or minimum wage laws—but this is apparently thought to be worth the price: a large group of moderately stigmatized people is worse than a small group of highly stigmatized people.

In sum, the long-term trends away from the self-conscious use of stigma to solve the incentive problem, away from reliance on in-kind benefits, and toward substitutes for welfare handouts, all might be tied to an increasing discomfort with the use of stigma as a tool of social policy.[8] However, I do not want to claim too much about the relevance of this history for current debates about public policy. Neither the history nor the model that it is supposed to illustrate provide a complete justification of minimum wage laws and similar laws. They do, however, provide a clue to their popularity, which the economist finds so puzzling both on normative and on positive grounds.[9]

11

Incommensurability, Commodification, and Money

Consider the following cases:

A person cancels lunch with a friend and feels bad, but he could not properly offer the friend $20 as compensation. Friendship and money are not commensurable (Sunstein 1994, p. 785).

An employer could offer an employee an extra $1000 to do work that would require an absence from home for a month, but could not, without insulting the employee, offer to pay the employee $1000 to persuade her to work away from home for a month. Family relations and money are not commensurable (Sunstein 1994, p. 785; Raz 1986, pp. 348–349).

A beautiful mountain may strike awe in the heart of an observer, but he would not say something like "this mountain is worth a million dollars." Environmental wonders and money are not commensurable (Sunstein 1994, p. 786; Raz 1986, pp. 348–349).

One should not offer to pay a neighbor $20 to mow one's lawn, or offer to pay a person for sex. Neighborly and intimate relationships are not commensurable with money (Sunstein 1994, pp. 786–787).

It is insulting to suggest that an actor or teacher's worth is equal to her pay or any amount of monetary compensation (Sunstein 1994, p. 788).

Sunstein (1994) and others have appealed to illustrations such as these to support the view that options are often incommensurable. The incommensurability of options is then presented as an objection to welfare economics, which typically assumes that options can be valued along a common metric whose units are dollars.

Although the incommensurability intuitions are powerful, they run up against an uncomfortable fact, which is that people often trade incommensurable goods and dollars. Sunstein mentions the "desperately poor person" who might abandon a friend for a sum of money, but he is altogether too delicate. Lawyers, businessmen, movie moguls, politicians, and even ordinary people constrained by time and resources drop friends when the costs of maintaining the friendships become too high. People routinely risk or violate family relationships by accepting attractive job offers in distant locations. Individuals express valuations of the environment when they choose detergents, buy paper products, recycle newspapers but not bottles, purchase large houses rather than small apartments, and litter. People pay for sex or for companionship, sometimes overtly and other times in carefully disguised manners: payment occurs in the form of gifts. Artists and teachers become investment bankers when their pay falls below a certain level. Workers accept premiums for risks, and government agencies use these risk premiums in order to calculate the costs and benefits of regulations.

The puzzle, then, is why people say, and even believe, that certain options are incommensurable, and sometimes act in ways designed to maintain the pretense of their incommensurability, while at the same time acting as though those goods were commensurable. Sunstein seizes one horn of the dilemma by arguing that the plurality of values and the incommensurability of options are fundamental, and that a theory of choice that accounts for them can explain people's actions. I will seize the other horn of the dilemma and argue that "incommensurability claims," as I call them, emerge in an equilibrium in which people rationally seek partners for the purpose of obtaining cooperative gains.

A Model of Principled Behavior

The incommensurability thesis holds that people cannot always value options along a common metric that is normatively justified. Most advocates of this thesis argue that people can choose among options but that the choice depends on qualitative differences between options that cannot be reduced to vectors on a single dimension of evaluation.[1] Although these advocates present this thesis as a claim about the morality of choice, I will treat it instead as a sociological fact: the tendency of many people to refuse to make tradeoffs, in everyday life, that are said to be demanded by the theory of rational choice. This chapter argues that this behavior is consistent with the cooperation model described in Chapter 2.[2]

My argument is that although the incommensurability thesis often describes people's *representations* about themselves, it does not describe their actual *behavior,* that is, the choices they make in everyday life, although their representations sometimes influence their behavior. People rationally make incommensurability claims in order to obtain strategic advantages in their interactions with others. Incommensurability claims do not reflect people's interests and values; they conceal them.

In the cooperation model presented in Chapter 2, a person's reputation refers to the beliefs other people have about his cooperativeness (or discount rate). This interpretation suggests that reputation is a continuous variable, and indeed we often speak this way. We say that X has a better reputation than Y, that X's reputation has improved as a result of his handling of a situation, and so on. But we also often speak as though having a reputation means being "principled" in a certain sense (Schelling 1960, p. 34). In ordinary talk people often say that X is principled or unprincipled, as though this aspect of reputation were a dichotomous variable. One cannot be a little bit principled. How are we to explain this idea of "principledness?"

The common-sense answer is that an unprincipled person acts out of self-interest and a principled person does not, instead conforming to the rules of morality or to the demands of altruism. This distinction, however, is not helpful. One often observes people who are considered to be principled violating their principles when the gains are sufficiently high. In addition, altruism is different from principled behavior. The parent who violates every rule of morality in order to help her child get ahead is an altruist but not a principled person. Conversely, a person who makes charitable donations might be motivated by principle rather than by altruism. The question is, then, How can one explain principled behavior without abandoning the standard premises of rational choice?

If we rely on the cooperation model, we do not abandon these premises, but we do not, at first sight, explain the dichotomous nature of principled behavior, either. The cooperation model seems to imply that if people believe that a person belongs to the good type, then they believe that he is principled; if people believe that a person belongs to the bad type, then they believe that he is unprincipled. The problem with this argument is that it does not explain the choppiness or discontinuity that characterizes principled behavior, so much as the plain fact that some people cooperate more than others. Consider three people: A, B, and C. A has a very low discount rate, B has an ordinary discount rate, and C has a high discount rate. One might therefore observe that, in ordinary relationships, A never cheats, B rarely cheats, and C frequently cheats. The question is why people would say that A is principled and C is not, rather than saying that A cares more about future payoffs than C does. And why is it that A would never say, "I have never cheated, but only because a sufficiently valuable opportunity has not yet presented itself"? Why would B and C insist that they are principled, and try to explain away or rationalize their defections rather than admitting them?

One can solve these problems by focusing on the problem of imperfect information. Suppose that X is trying to decide whether to enter a relationship with Y. X has no past experience with Y, so X cannot infer Y's discount rate from his own observations of Y's behavior. He knows that Y has a husband, several commercial partners, many friends, and he has heard that she has never cheated on her husband or her friends, but that many of her commercial partners complain that she cuts corners. X's problem is that he does not know how reliable any of the information—positive or negative—is. Maybe Y has cheated on her husband but is discreet; maybe Y has not cheated on commercial partners but they are envious of her success and want to draw away her customers. If other desirable partners are available, X might refuse to enter a relationship with Y, despite her many attractions.

To enhance her appeal to X and others, then, Y sends signals: she gives gifts, shuns the people one "should" shun, wears the clothes that one "should" wear, speaks the way one "should" speak, and so on. These signals, however, will not always succeed in revealing one's type, and they are likely not to succeed when signals are too noisy. Consider a well-known model of firm behavior, in which the good types must charge high prices in order to recover the cost of manufacturing high-quality goods (Klein and Leffler 1981). A bad type offers a low price, claiming that the low price is due to its more efficient methods of production, not to the low quality of its merchandise. To attract customers, the good types signal by advertising. The good types can distinguish themselves from the bad types only if customers believe that only good types can afford expensive advertising. However, customers might not believe that firms that purchase expensive advertisements produce better goods. The good-type firms might try to implant this belief in the minds of customers by saying that their goods are better than those of their competitors, rather than using empty advertisements. But customers might not believe this statement, or they might not hear and remember it amid the cacophony of competing advertisements. Even if customers believe that firms that use expensive advertisements are more likely to produce better goods, they might not be able to distinguish expensive and not so expensive advertisements. Then, a bad firm that purchases not so expensive advertisements is indistinguishable from a good firm; the good firm may then mimic the bad firm's advertising strategy in order to save money.

The result is that the good and the bad firm will say in their not so expensive advertisements that they always deliver high-quality goods. There are two theories for why they will make this statement. First, they must say *something* in order to persuade buyers that the advertisement shows that they are good types, and the natural thing to say is that they are good types. They will say "I never cheat" (that is, "my goods are high-quality"), rather than "I usually never cheat" or "I cheat 1 percent of the time," because the claim that one never cheats is focal (Schelling 1960, p. 57). A firm that wants to persuade buyers that it is reliable makes an absolute claim of reliability (100 percent reliable), rather than a partial claim (98 percent or 97.5 percent reliable), even if the partial claim is more honest, because people do not differentiate among shades of reliability and will interpret anything less than a claim of absolute reliability as an admission of substantial unreliability. Given that in signaling models, *any* action with the right cost structure can be a signal, the most natural signal in the advertising context is a claim that the goods are always high-quality.

The second theory for why all kinds of firms will claim never to cheat is that some buyers might be credulous (by which I mean poorly informed), even if most buyers are not. If these buyers accept statements at face value, then they will buy from the firm that makes the stronger claim that it does not cheat. Any firm that honestly said that it cheats 10 percent of the time will lose these buyers to a firm that said that it cheats 9 percent of the time. Firms will bid each other down to 0 percent in order to avoid losing buyers to each other.[3] The same result is obtained by assuming that none of the buyers is poorly informed but that a tiny fraction of firms really are principled.

To be clear about why people act this way, put yourself in the shoes of the buyer in the latter example. If the seller tells you that he never cheats, then you might not believe him, but because all the other sellers say the same thing, this statement gives you no basis for rejecting this seller in favor of some other seller. But if the seller tells you that he sometimes cheats, while all other sellers deny ever cheating, you might as well take your chances with another seller. One might argue that the buyer should admire the seller who admits that he cheats: this seller is, at least, honest. But someone who cheats is, by definition, dishonest. If he says that he cheats, you have no reason to believe that he is more honest than people who deny that they cheat. It is possible that someone in the group of people who deny that they cheat really is honest; it is impossible that the self-proclaimed cheater is honest.[4]

Even if you are so cynical that you would never believe absolute or extreme claims, you are not likely to treat differently the seller who makes absolute claims and the seller who does not. You gain nothing by treating them differently. (Indeed, if you and others tried to treat them differently, they would conform their behavior to your expectations without being any less likely to cheat you. If customers trust most those firms that announce that they are 50 percent honest, all firms will announce that they are 50 percent honest!) But if some people are credulous, sellers benefit by making absolute claims. These claims cost them nothing. They do not lose the cynics' business, and they gain, or avoid losing, the business of the credulous people.

In sum, firms are forced to make exaggerated claims about the purity of their motives and the quality of their products. These exaggerated claims never to cheat can be described as claims to be principled. The claim that one never cheats is a claim that one is principled; the claim that one occasionally cheats is not.

The analysis does not only apply to relationships between firms and consumers. It applies to any cooperative relationship. The bad or worse types will claim to be principled either because of the possibility that someone will believe them or because principledness is focal. The good types have to follow the bad types, or risk being mistaken as bad types. Returning to the earlier examples, the buyer will say, "I never pay more than rock-bottom price." He will never say, "I never pay more than rock-bottom price unless I'm really desperate for the goods," even though the latter statement is surely more accurate than the former. In the second example, the seller will say, "I never deliver late," rather than, "I deliver late only when I am having serious labor problems." In social relationships, people say to their friends and lovers, "I will never betray you," or "you can count on me," rather than, "you can count on me unless the cost of doing so exceeds a certain threshold."

This is not an argument that a seller would never say that it will not have labor problems that might interfere with its ability to perform. Some sellers do say that, but that is not the same as saying that one will cheat. When a seller informs the buyer of the possibility of labor problems interfering with delivery, and the buyer nonetheless accepts the seller's offer, no doubt with a price adjustment that reflects this risk, then failure to deliver because of labor problems does not count as cheating. The value-maximizing obligations that emerge in a relationship can be rigid or flexible, depending on context.

Two further points should be made. First, a person who claims to be principled is not speaking truthfully, but his dishonesty is not harmful because no one is deceived. As mentioned before, a rational person will sacrifice his reputation when the gains are sufficiently high. But a principled person is one who claims never to make such a tradeoff. The person who claims to be principled does not say: "I will not cheat unless the gains exceed reputational losses." He says: "I will never cheat." People are driven to extreme statements by their focal nature and by the need to match the exaggerated claims of bad types. Thus, all people claim to be principled but they are believed by no one except for credulous people (such as children, whose discovery of this fact about the adult world is the beginning of their disillusionment). An air of corruption hangs over every bargaining table.

Second, good types try to distinguish themselves from bad types by behaving consistently with their words as much as possible and making claims that can be checked out. The claim that one is principled cannot be disconfirmed by the addressee until it is too late—until the speaker has cheated. But the speaker might overcome the problem of the cheapness of talk by integrating claims about his principledness into a general theory about himself. Consider the following common statements. "I feel terribly guilty when I tell a lie." "My parents taught me never to take advantage of other people." "God punishes those who tell lies." "In the army we were taught always to tell the truth." Statements like these are more plausible than acontextual assertions of truthfulness, because the speaker reveals other things about himself that can then be used to verify his truthfulness. If the first speaker later plays a nasty practical joke, the second speaker turns out to be an orphan, the third an atheist, and the fourth a draft dodger, then the listener should start taking precautions.

In their effort to distinguish themselves from bad types, some good types engage in speech and behavior in which they would not engage but for the problem of distinguishing themselves. Note that good types are driven not only to distort their speech but also to distort their behavior. A non-religious person might claim that he is principled, then provide as evidence the fact that he spends a lot of time at church. If spending long hours at church is sufficiently costly and differentially costly, the person will distinguish himself as a good type by making this claim—but only if his claim can be checked out. If others observe that he does not go to church, he will be exposed as a bad type. The person goes to church, then, as a way of showing that he is principled, not because he seeks religious solace. His investment in this signal enables partners to cooperate with him to a greater extent than with bad types who cannot afford the signal, and in this sense serves to enhance cooperation. But the person is still not *principled,* because he will cheat if an exogenous change modifies his payoffs by a sufficient amount.

Before moving on to the issue of incommensurability, let me summarize the argument so far and respond to an important objection. The argument is that people acting in their rational self-interest face powerful incentives to claim to be principled in the ordinary sense of the word, that is, never to cheat in the cooperative relationships to which they belong. Having claimed to be principled, people must act consistently with the demands of principle, lest they be immediately perceived as bad types. At the same time, people will cheat when the short-term gains outweigh the long-term reputational costs. Given the focal character of absolute statements, and the existence of some people who are credulous, one has a strong incentive not to admit the truth, and to claim, to the contrary, that one never cheats. Since in equilibrium no person admits that he would ever cheat, the statement that one does cheat will cause people to infer that one belongs to the bad type.

It is worth emphasizing that because sometimes a person's principled claims will constrain his behavior, the person will cheat less often (though he will cheat if the payoff is high enough) than he would if he did not make principled claims. Therefore, claims to be principled actually may produce social benefits—by reducing the amount of cheating—even though they do not ensure or reflect principled behavior.

The objection to the theory is that it treats people like "rational fools," in Sen's (1977) words. Sen argued influentially that economics assumes that people are rational calculators, whereas in fact people are often motivated by principle, or what he calls "commitment." My theory shows why people act *as if* they were motivated by principle, and it therefore addresses Sen's criticism to the extent it is intended as a critique of the explanatory ambitions of rational choice theory. To be sure, Sen's argument was also that the methodology of economics misses an important side of human motivation, phenomenologically speaking. The problem with Sen's argument, however, is that simply assuming that people act out of principle *and* rational calculation gives one less methodological purchase than the ordinary rational choice assumptions do, without, as far as I can tell, compensating for this loss by producing a methodological gain.[5]

Incommensurability

I have used the idea of being principled in a broad sense, pretty much as the equivalent of being a person who does not cheat in relationships, a cooperator. People claim to be principled because others infer that people who do not make such claims are bad types. Similarly, people claim that options are incommensurable, because others infer that people who do not make incommensurability claims are bad types.

To understand this point, suppose that people believe most things a person says about himself, and one person wants people to enter cooperative relationships with him. What sort of things might that person say? Recall that a person is an attractive cooperative partner if he has a low discount rate, because a person with a low discount rate is relatively unlikely to cheat in a game characterized by indefinitely repeated transactions. But the low discount rate is not a sufficient condition for cooperation. It must also be the case that the person gains, in any given round, relatively little by cheating when the other person cooperates. So in order to attract cooperative partners, a person will say that (1) he has a low discount rate and (2) his outside opportunities are not valuable.

People make these claims in complex ways. First, they claim to be mature rather than immature, self-disciplined rather than impulsive, strong rather than weak. All of these claims are about discount rate. Second, people claim that if they cheated, they would feel guilty or ashamed, they would burn in hell, or they would not be trusted by other people. All of these claims are about outside opportunities. Third, people claim that they are principled or are honorable, or that they do not engage in rational or strategic behavior. These claims constitute a denial that one is even capable of engaging in the sort of calculations that make strategic behavior a possibility in the first place.

When a person makes an incommensurability claim, he is making a claim of either the second or the third kind. One incommensurability claim is the claim that the benefits from cooperating cannot be compared with the benefits from cheating. To admit that the two kinds of benefits are comparable would be to say that one will choose the action (cooperation or cheating) that provides the higher benefit. Denial of the possibility of comparison is a denial that the speaker can engage in strategic behavior and, by implication, a guarantee that the speaker can only cooperate. The incommensurability claim is the claim that there is no question of a tradeoff: the benefit from cooperation is "differently valuable" (Sunstein 1994) from the benefit from cheating.

The other incommensurability claim is that the benefits from cooperating are infinitely high. If the benefits from cooperation are infinitely high, whereas (by implication) the benefits from cheating are finite if sometimes very high, even a rational maximizer will never cheat, if that means that in a later round the gains from cooperation will be lost. The incommensurability claim is again that there is no question of tradeoff, but this time because the benefits of cooperation are so high.

These two ways of claiming to be principled track a dispute between Sunstein and Regan. Regan (1989) argues that the incommensurability claim implies that the speaker places an infinite value on something. Sunstein (1994, p. 813) argues that the incommensurability claim implies that the speaker places a different kind of value on something. Because people make both kinds of claims, and both kinds of claims serve the same general strategic purpose, for my purpose there is no need to make such a distinction.

As an example, consider a person who refuses to answer a cost-benefit survey about the value of a mountain, or a mountain view, claiming that no amount of money can compensate him for obstruction of that view. The commitment to the environment serves as a signal, in this person's community, of one's loyalty; anyone who says that the mountain view can be traded off against something else would be considered a bad type. To show that he is not a bad type, a person would deny the possibility of such a tradeoff. This claim could be interpreted as an assertion that the mountain view has an infinite value, or as an assertion that comparisons cannot be made—it does not matter. Both claims serve the strategic purpose of signaling one's loyalty to the relevant community.

Why isn't the simpler interpretation that the person places an infinite or incommensurable value on the mountain view?[6] The answer is that except in unusual circumstances people do not act as if they placed an infinite value on anything: people routinely make tradeoffs in their lives. People sacrifice mountain views so that they can move to a location where there are better jobs, better schools for their children, and more culture. When pressed, people admit that they would give up a mountain view if that meant saving lives, producing medicine, or reducing crime and unemployment. A person who places an infinite or incommensurable value on something *in fact* would give up everything he owned, every future opportunity, every good for other people, in order to maintain that thing. If there are any people like this, they are too few to have strong claims on the direction of public policy.

A response to this argument is that the person would not trade the mountain view for money but would trade it for other goods, such as the reduction of unemployment in the region. The mountain view and money are incommensurable, but the view and employment are comparable. But it is hard to see why, if a view and unemployment are comparable, they cannot both be compared to money, which serves as a metric. A sum of money has as much in common with a mountain view and as much in common with a reduction in unemployment as the latter two have with each other. The implicit assumption here is that the sum of money would be used for a self-interested, rather than other-regarding, purpose—*that* is why trading the mountain view for money would be objectionable, just as trading the mountain view for a private swimming pool would be objectionable. The reason that a person in the hypothetical community could *overtly* make a tradeoff between the mountain view and the reduction in employment, but not between either of these options and a sum of money or a private swimming pool, is that one obtains reputational gains by professing respect for the quality of life of others, not by demonstrating one's concern with one's self. Assuming that his statements are believed, a person who says that a mountain view can be sacrificed for a $1 million gain sends a more ambiguous signal than the person who says that a mountain view can be sacrificed only in order to reduce unemployment. The latter person is more clearly willing to incur a cost than the former, who, others might think, might expect to receive a share of the cash.

Another example is a professor who refuses to accept money to skip a faculty meeting but who would skip the faculty meeting in order to stay home with a sick child when childcare is not available. One might believe that these options are incommensurable. Preliminarily, note that it is not because it involves money that the monetary option is unacceptable. Most people would think that the professor could skip the faculty meeting if the government threatened to fine him $100,000 if he goes. The reason that the bribe is unacceptable and the threat and the childcare option are acceptable is that the implicit, value-maximizing obligations that emerge among faculty members allow one to skip meetings in order to meet serious personal obligations that conflict with them, or to avoid serious harm. These obligations do not allow one to satisfy predictable fluctuations in tastes or to seize predictable opportunities. This view is standard in contract theory. Nothing about the example suggests that such behavior is inconsistent with the choices of rational actors.

A related example is the person who refuses to trade time with his family for money but would trade time with his family for a better career opportunity. This person does not claim that time with his family has infinite value, but does claim that it cannot be compared with money or even with the better career opportunity, on a common metric. But it is hard to see what is accomplished by this argument. It is likely that trading time with one's family for money would count as cheating on the family relationship. Trading time with one's family for a better job opportunity would count as cheating in some families but not in others—it simply depends on what value-maximizing obligations emerge. Incommensurability arises because people promise not to trade off time with the family for anything. This absolutist claim is, again, more focal and more resistant to the competing exaggerations of bad types than a more honest and limited claim, such as, "I will not trade off time with my family except for something that is very important for me." The moral awkwardness of the tradeoff reflects the sense that one is caught in a game that denies its players the luxury of honesty.

As discussed above, this is not to say that incommensurability claims are empty talk. In fact, these claims put people in a bind. On the one hand, one cannot truly act consistently with them. On the other hand, if one does not act consistently with them, one may lose the reputation that they are intended to establish. Ordinary people shade their behavior so that violations of their incommensurability claims occur relatively infrequently, and occur only under extreme circumstances. But the problem torments politicians because their words and actions are scrutinized so carefully by the press. The politician who refuses to make incommensurability claims will be called a person lacking in vision and integrity. The politician who makes incommensurability claims will be called principled or rigid. When the latter is forced to compromise, he is then called slick. Ordinary people can avoid these problems as long as the incommensurability claim is consistent with rational action, that is, cooperation remains rational; when it does not, they cheat and try to conceal their opportunism behind casuistry.

A third example is the friend who cancels lunch. A person X who offers his friend Y $20 as compensation for a canceled lunch does something wrong. But what is wrong? Suppose that Y reasons as follows:

> If X offered me $20, then X must believe that I think that X's cancellation was a form of cheating that would cost me $20. If X thinks that I place a monetary value on this (minor) form of cheating, then X must think that I would also place a monetary value on more significant forms of cheating or betrayal. But that implies that X thinks that I would betray him if the payoff were sufficiently great. So by offering $20, X implies to me that he believes that I will betray him if the payoff is high enough. But if he thinks that I will betray him when I get the chance—that, for example, I will not come to his aid if something really bad happens to him—then I can hardly expect him to refrain from betraying me if the price is right. But if a friend can't do that, what are friends for?

But why doesn't Y say to herself, "X is no saint, but he still can be my friend; only saints place an infinite value on friendships." The answer is that absolutist claims are focal and resistant to the exaggerated claims of bad types. If X values Y's friendship at $A, then someone else will claim that he values Y's friendship at $A+1; then another person will claim that he values Y's friendship at $A+2; and so on. Only an infinite valuation cannot be trumped by an effort by a bad type to insist that he values the friendship a certain amount more. What frustrates the good type is that although by claiming to value the relationship at an infinite amount he avoids the inference that he belongs to the bad type, he is also unable to distinguish himself from the bad types and he has also been corrupted by his emulation of them.[7]

Incommensurability claims mask a frustration that many scholars feel toward money. They believe that money diminishes important practices, values, and relationships. Economists ridicule these claims, pointing out that money is just an index for valuing goods and services that we cannot avoid trading off. If money did not exist, we would engage in barter, but the sort of tradeoffs that offend incommensurabilists would still occur. There is nothing inherently bad about money. But the economists' response is not correct.

The problem with money is that by supplying a scale for ranking all options it exposes the strategic basis of cooperation, and by exposing the strategic basis of cooperation, it undermines it. Given the observable alternatives supplied by the market, the value a person places on a particular relationship can be estimated within a range. That person must exaggerate the value he places on the relationship in order to avoid the inference that he belongs to the bad type and to persuade his partner that he values the relationship at the upper end of the range, but the exaggeration is an embarrassment because everyone knows that it is an exaggeration. This is the source of both the stigma that is associated with money and the market, and the feeling of alienation that the market economy is said to produce.[8] The stigma reflects the same dynamics that give rise to incommensurability claims: people stigmatize others who too overtly desire money in order to represent, by comparison, their own immunity from being bought off or tempted to defect by a higher payoff. This is why people must constantly deny that they are motivated by money even as they rush around accumulating it. The feeling of alienation is the product of the dissonance between one's behavior, which inevitably reflects tradeoffs, and the public denial of this fact.

The analysis is the same for neighborly, intimate, work, and political relationships. These are all cooperative ventures in which it is important to signal that one is not a cheater. Parties to the relationship continuously send each other signals of their loyalty, precisely because they know—and they know that their partners know—that it may be worthwhile for them to break off the relationship. One way to signal loyalty is to say that the relationship is incommensurable with other options. Few people really believe this, but no one can express this skepticism openly without risking the end of the relationship.

As we saw earlier, the claim that environmental goods are of infinite or incommensurable value also illustrates the use of incommensurability claims to signal that one will not cheat. People who are in environmental groups (defined broadly, to include people who are in relationships of any sort with others who also value the environment a great deal) show their loyalty to other members by saying that they place infinite or incommensurable value on the goals to which the group is devoted. People do this in order to persuade others that they do not belong to the bad type. This logic can be extended to other activities. Teachers tell each other that teaching is of great value in order to persuade each other that they are loyal to the group endeavor. So do research scholars, doctors, lawyers, actors, artists, and no doubt everyone else. Telling a co-worker that "we are paid what we are worth" is not going to endear you to him; telling him that "we are all underpaid, and no one appreciates the value of what we do" might. The person who makes the first statement cannot be trusted in, say, the formation of a union; the person who makes the second statement at least might be trustworthy.

An objection to this discussion of principledness is that people are not as cynical as the argument implies. The argument, however, does not claim that people are cynical, or that they make incommensurability claims cynically. Many people believe their incommensurability claims. One can sincerely make incommensurability claims as long as circumstances do not provide a valuable opportunity to cheat. As long as the demands of interest and of ideology converge, there is no sense of cognitive dissonance that must be resolved, at least for an unreflective person. When they conflict, the usual response is to rationalize one's defection, that is, to revise one's story about oneself or one's action in a way that is believable and that preserves one's reputation or self-image as a principled person. For example: "that was not really cheating, because the other person was about to cheat me, anyway"; or "everyone has a lapse now and then, and this won't happen again."[9]

Normative Implications

It would be convenient if the theory I have described implied that the government should simply ignore incommensurability claims when determining policy, but the story is more complicated than that. Imagine that the government must decide whether to preserve a small patch of forest or to allow companies to exploit it. The benefit from exploitation is the creation of jobs and the production of important drugs. The cost is the elimination of a piece of the environment. When asked, people say that this piece of the environment has infinite value, or cannot be reduced to a sum of money, or cannot be compared with other goods, and in any event refuse to assign a monetary value to it. What should the government do?

The simple response is to say that the government should ignore the responses and estimate the amount of money that people would trade off in order to preserve the piece of forest. Because these people do not really value the project at negative infinity and do not really believe that the losses and the gains are incommensurable, their survey answers should carry little weight with the government.

The first problem with the simple response is that it ignores the important role that incommensurability claims play in social, commercial, and political life. Even though incommensurability claims do not reflect true valuations, they are often valuable as forms of behavior. They help maintain cooperative relationships, and cooperative relationships are valuable as long as they do not injure third parties. If the state could forbid people to make incommensurability claims, this prohibition could (though not necessarily) make people worse off, because it would be harder for people to signal their suitability for a cooperative relationship. Of course, the state could not forbid people to make such claims; but even ignoring such claims might discredit the claims in a way that reduces their utility for strategic purposes. Indeed, from a utilitarian perspective the incommensurability problem puts the government in a bind. If the government takes incommensurability claims seriously, it would fail to make tradeoffs that increase social welfare. But if it discounts them, it might interfere with people's efforts to engage in cooperative ventures.

Still, one should distinguish the existence of incommensurability and other principled claims from what these claims are *about.* The government cannot prevent people from making *any* incommensurability claim, and even if it could, there is no reason to believe that the new equilibrium would be characterized by greater social wealth. But some incommensurability claims are more socially costly than others. People for whom ethnic purity is a matter of principle might be superb cooperators within their own ethnic group. These people claim that they will not cheat on members of their own group in order to obtain benefits from outsiders. This is an incommensurability claim because it denies that any amount of outside benefit will break the bonds of ethnic loyalty. The effectiveness of this claim at promoting cooperation can be seen in the history of nationalist movements. In the United States, the claim has frequently been advanced by whites as a way of strengthening their bonds to each other. Yet clearly this kind of incommensurability claim deserves no respect from the government, for the advancement of this claim commits the speaker to harming third parties. A similar argument can be made about religious incommensurability claims. By comparison, incommensurability claims about one's family and friends are relatively, though not completely, benign.

Incommensurability claims about the environment present similar difficulties. Environmental protection is a principle around which people rally. They make incommensurability claims about the environment in order to show their commitment to like-minded people. Environmental incommensurability claims, unlike ethnic incommensurability claims, do not commit the speaker to behavior that directly harms third parties. The problem with these claims is that, if acted on, they will indirectly injure third parties—people will be denied jobs, medicine, and inexpensive food because of restrictions on development.[10] I am not arguing that development is necessarily good; I am arguing that incommensurability claims interfere with efforts to evaluate policy according to its contribution to the welfare of those affected, and that this interference is not justified by a vindication of authentic moral values. The interference is, rather, a side effect of the fact that certain political positions have become symbolic of a person's trustworthiness within certain groups. Similar comments can be made about incommensurability claims about bodily integrity, including the claim that people should not be permitted to sell their organs, and the claim that women should not be permitted to bind themselves to surrogacy contracts, whatever the benefit to them.[11]

The second problem with the simple response—that the government should just ignore or discount incommensurability claims—is that it is paternalistic. On what grounds should the government ignore the expressed views of its citizens? A simple answer is that the government should ignore these views when they do not reflect citizens' interests and values. When citizens make incommensurability claims in order to avoid reputational penalties, preferences will be concealed, and preference-respecting policies will not reflect them (compare Kuran 1995, Loury 1994). Under these circumstances the government should not defer to expressed preferences.

This is a theory of ideology, which was hinted at in Chapter 7. An ideology consists of propositions which people publicly endorse in order to show that they are good types. Public endorsements of these propositions count as signals of loyalty because people infer that those who reject these propositions are bad types. Ideologies are exaggerated and simplified versions of the world views that people actually hold. The nuance and blurriness of people's nonideological views about the world disappear in the ideological version, because nuanced positions do not serve well as signals, whereas extreme positions do. Ideologies are rigid: they rise suddenly and collapse violently. World views are flexible: they shift and evolve gradually. The rigidity of ideologies and the discontinuity of ideological change result from the self-reinforcing nature of signaling equilibria. As long as everyone takes the public endorsement of certain positions as evidence of loyalty, people will be reluctant to disagree, or even to express views that are a shade different; but once changes in circumstances increase or decrease the cost of deviation by a sufficient amount, the equilibrium collapses (Kuran 1995).

The role of the critic is to identify those popular propositions that are arbitrary focal points of loyalty and that do not reflect people's authentic interests. Critical theory seeks to identify distortions in the political process that undermine political representation. One such distortion is the equilibrium that results because people have strong incentives to signal their desire to enter and enhance relationships with each other. Because people do not want to offend their partners, they will not make certain criticisms which, if followed, would make everyone or most people better off. A purpose of critical theory, then, is to describe conditions under which such an equilibrium results, and to criticize the statements and actions that serve as signals in that equilibrium. Habermas' (1984) proposal that we imagine an ideal conversation in which people are forced to provide honest defenses to any claim they advance is an approach to this problem. In such a conversation, one would have to admit that one supports a particular policy in order to signal one's loyalty; thus, one's support for that policy can be discounted and one's true preferences revealed.

The use of signaling models to disclose ideological distortions in public debate falls short of the ambitions of Habermas and other critical theorists. It would not show whether any non-reputational preferences are ethically objectionable. But so ambitious a version of critical theory runs up against serious conceptual barriers. The marriage of signaling theory and critical theory avoids these problems while also having a great deal of purchase. It allows identification of areas of life and politics in which the perceived importance of signaling one's type overwhelms the incentives to speak or act truthfully, resulting in equilibriums in which non-reputational preferences are not satisfied. The critical theorist would have the role of persuading people that they would be better off if they acted and believed differently: his or her "imaginative reconstruction" would propose a new focal point around which behavior would shift.[12] This already occurs, of course. It occurs when people struggle to persuade others that costly signaling, like dowry competitions and conspicuous consumption, does not serve their interests. It occurs when norm entrepreneurs propose collective action—because that is always what is needed to escape a bad equilibrium—in order to change signals. The most prominent examples come from the civil rights, feminism, and gay rights movements, and religious revivalism; others include attempts to escape militaristic patriotism (during the Vietnam War) or to provoke it (at the outset of World War II), and the recurrent attacks on materialism and conspicuous consumption.

This leads to the third problem with the simple response. There is a vast gulf between the normative project of identifying distortions in people's political views and the political project of forcing people to change their views. Incommensurability claims present a political problem. On the one hand, the government cannot take them seriously, because then it could not make the tradeoffs that governments have to make all the time. On the other hand, if the government does not take them seriously, then people will press their incommensurability claims against the government—again, as a way of showing each other that they belong to the good type. This conflict is played out over and over on the stage of American politics. Economists, technocrats, and pragmatists trade off, balance, and compromise. Religious figures, moral leaders, and ideologues respond by asserting the impossibility of compromise over matters of principle. Politicians, in turn, are forced to take positions on these issues, as a way of signaling that they, too, are principled. American politics mirrors the conflicts faced by ordinary people: on the surface we see furious controversies over flag desecration, abortion, gun control, affirmative action, and other issues on which all politicians are required to have a firm and principled position. Beneath the surface we see frequent compromise—by the same people—about bridges and highways, taxes, and intricate regulatory statutes that lack the national salience of the other issues.

The *conceptual* problems identified by the incommensurabilists dissolve. People who take extreme positions on religion, free speech, and environmental protection are signaling to relevant communities that they can be trusted. They take extreme positions, rather than merely strong positions, because they fear being mimicked by bad types who claim to be more deeply committed than they are, and because extreme positions are more focal than nuanced views.

But the *political* problem remains. People will continue making incommensurability claims because of the pressure to signal trustworthiness. People will resist attempts by the government to discount incommensurability claims; and, indeed, politicians and other government officials will not want to try too hard to discount them, lest people infer that they are bad types. Still, the political problem should be distinguished from the conceptual problem. We care about incommensurability claims because they inevitably emerge in social life; but we should not concede that they reflect deeper values.

12

Autonomy, Privacy, and Community

Wherever interaction is continual, dense, and valuable to participants, distinctive patterns of behavior emerge. When deviations from these patterns provoke nonlegal sanctions, we say that the patterns are social norms. Social norms both create options and suppress them. People who leave a busy city and set up a commune in the country believe that the norms that emerge in their community will create a new and valuable option for themselves and their children—the option to live an uncorrupted life. But their children may believe that the creation of this new community has removed the most valuable option from their lives—the option to obtain a secular education and pursue a worldly career.

The children, as they grow up in the commune, or any small town, or an ethnically or religiously homogenous enclave, or any other tight-knit community, might flourish, but might not. Those whose tastes and values separate them from the dominant group will feel themselves oppressed by the weight of community sentiment. If they consequently violate social norms, they find themselves shunned, but maybe also trapped and unable to leave. In earlier chapters, these problems were examined from a welfarist perspective. This chapter examines them from the perspective of autonomy. Nonlegal enforcement of social norms may have objectionable effects on individual autonomy, even in the eyes of those who are relatively happy in their community and strong believers in its values. The pathologies of nonlegal enforcement justify two broad legal interventions in community life: the protection of privacy and the rule of law.

Autonomy

Autonomy is an important value, but because it is more difficult to understand than efficiency and distributive justice, I must begin with a brief discussion of what autonomy means. Consider two examples of people who *lack* autonomy (quoted from Raz 1986, pp. 373–74).

> *The Man in the Pit.* A person falls down a pit and remains there for the rest of his life, unable to climb out or to summon help. There is just enough food to keep him alive without (after he gets used to it) any suffering. He can do nothing much, not even move much. His choices are confined to whether to eat now or a little later, whether to sleep now or a little later, whether to scratch his left ear or not.
>
> *The Hounded Woman.* A person finds herself on a small desert island. She shares the island with a fierce carnivorous animal which perpetually hunts for her. Her mental stamina, her intellectual ingenuity, her will power and her physical resources are taxed to their limits by her struggle to remain alive. She never has a chance to do or even to think of anything other than how to escape from the beast.

The Man in the Pit and the Hounded Woman have some freedom, but not much. The options available to them are trivial. They are trivial for the Man in the Pit because no choice he makes affects important aspects of his life. They are trivial for the Hounded Woman because every choice is dominated by the life-preserving option. What is missing from their lives is *autonomy.*

What is autonomy? It is useful to begin by comparing autonomy to utility. Economists say that when an individual has a choice between A, B, and C, and prefers A to B to C, this person maximizes his utility by choosing A over B and C. Utility refers, then, to preference-satisfaction. Notice that utility is subjective: the person's utility depends on his satisfying his preference for A, not on an objective view that A is superior to B and C.

As a first approximation, one might say that autonomy refers to the person's capacity to choose between A, B, and C. This might make autonomy seem redundant with utility: if a person with the capacity to choose between A, B, and C always chooses A, why should anyone care that he has the capacity to choose? Clearly, it would violate the person's autonomy and at the same time decrease his utility to take away A and give him C. The purpose of the distinction becomes clearer when one asks what happens if the state takes away B and C, leaving only A, or even orders the person to choose A. From the perspective of utility-maximization, this act does not offend one, since the person still gets his first choice. From the perspective of autonomy, however, one is troubled: the person gets what he prefers, but he does not get a *choice.* Putting this point differently, a person's utility increases, even though he does not exercise autonomy, when the state provides him a good that he prefers but that he does not choose. A person exercises autonomy, even though his utility does not increase, when the state provides an alternative to a good that he currently enjoys and he rejects the alternative because he prefers the good.

To say that a person exercises choice over a range of options is not necessarily to say that he is an autonomous person, any more than saying that a person who obtains a desired good is a happy person. The Man in the Pit has choices about when to scratch his ear, but he is not an autonomous person; he will scratch his ear if he prefers scratching it to doing something else, but he is not a happy person. Raz argues that the autonomous person has three characteristics. First, he has sufficient intelligence, health, imagination, and emotional stability to be able to make meaningful choices. Second, he has an adequate range of options—as mentioned above—that include activities that are both trivial and significant, long-term and short-term, and so on. Third, he is not coerced or manipulated with respect to any of those options, even if this would leave him an adequate range of options under the second criterion.

These criteria are vague, but that is only because very different lives are consistent with the ideal of autonomy. A person who enters a monastery or the army is an autonomous person so long as he has some alternatives (for example, working in a factory) and is not coerced (for example, by his family or the police). A person who moves from job to job may have autonomy; so might a person who remains unemployed in a bad market, as long as he exercises sufficient choice in other aspects of his life. A common sort of person who does not have autonomy is one in a highly traditional society, in which his choices with respect to employment and marriage are made for him. A person in jail does not have autonomy, though we might say that he is an autonomous person if his imprisonment is a consequence of his freely chosen life of crime. Although some people might think that autonomy has a deeper meaning—that it requires, for example, that a person develop a "life plan" and follow through on it—the definition used here is surely a necessary condition of the deeper definition, and is adequate for present purposes.[1]

Increasing individual autonomy. These considerations raise the question of whether it is possible for the state to increase the autonomy of a person. In some respects, the answer is obviously yes. It can release a person from prison. But the obvious answers do not help the modern developed state, which faces complex questions about the optimal allocation of resources. The problem is that the state does not necessarily increase a person's autonomy merely by increasing his range of options. If a person already has an adequate range of options, an additional one makes no difference. If the government creates a new job, the mere existence of this new job does not increase my autonomy if I already have an adequate range of options with respect to my working life.

The welfare system provides another example. Suppose there is no welfare system. I am so poor that I work all the time to make ends meet, and have no options for other jobs, or other uses of my time. I am like the Hounded Woman. Now create a welfare system. The security provided by the welfare system gives me the option to quit, and to use my free time to invest in skills for the purpose of getting a better job. Or it may allow me to work less hard, because I no longer fear getting fired quite as much. I now appear to have more autonomy. But the availability of the cash transfer also in some respects reduces my autonomy. I now no longer have the option of being entirely (or even mostly) self-supporting. Of course, I could turn down the cash transfer, and this is a *new* option. But my life has nonetheless in a substantial respect changed: I always know that I can turn to the government if I have problems, so I cannot be entirely self-supporting. It is a mistake, then, to say that the welfare law increases or reduces my options. It *changes* my options; and whether this change increases or reduces my autonomy depends on how we evaluate the options that are lost and gained.

One can identify five ways in which the state can increase or decrease the autonomy of citizens. The first three ways are obvious, and I mention them only for completeness.

Poverty. People living in abject poverty have, like the Hounded Woman, few options, and must choose whatever work that will support them. Notice, however, that there are great difficulties in determining how much poverty sufficiently impinges on autonomy. Even very poor people have some choice among jobs, and can get married, raise families, and enjoy leisure; if we limited our concern to people who lack these choices, then poverty would not be a relevant concern in most developed countries.

Ignorance. An ignorant person does not have the capacity to make valid choices. Again, we have a problem about how much ignorance is necessary before we believe that a person is incapable of having a sufficient amount of autonomy. Some people might say that only severely mentally disabled people lack the capacity to have autonomous lives; others might argue that anyone lacks that capacity if he has not been exposed to a sufficient number of views about what it means to lead a good life.

State (and private) coercion. The state can restrict autonomy in straightforward ways. It can throw people in jail, or threaten them with jail if they do not follow the approved path of conduct. If the state controls people's choice of employment, it limits their autonomy both by restricting options and violating independence. If the state taxes people a little, it violates their autonomy in a minimal way or perhaps not at all; if it taxes people a lot, or in a way that is calculated to cause them to engage in approved activities, then it violates their autonomy in a significant way.

Now let us turn to the more interesting fourth and fifth ways that a state can affect people's autonomy.

Social Influence, State Influence, and Autonomy

Consider the following characters.

> *The Pampered Prisoner.* A man lives in a giant building, entirely alone, and can never leave. The building has many rooms, each of which contains all the resources necessary to develop any skill or ability. One room contains a gymnasium; another room contains a vast library; a third room contains thousands of botanical specimens. The man thus has choices among a variety of significant life plans. He is also supplied with food (but of course he can refuse it and grow his own food in the courtyard and cook it in the kitchen).
>
> *The Hounded Villager.* A woman lives in a small village and cannot leave it. She knows everyone in the village intimately, and they know her intimately. She legally has the choice of the kind of education to obtain, whom to marry, what religion to observe, what job to take, and where to live. But if she obtains the wrong kind of education, marries the wrong person, observes the wrong religion, takes the wrong job, and lives in the wrong place (where "wrong" means in violation of the expectations of the other villagers), the villagers will ostracize her—that is, ignore her, refuse to deal with her, avoid her.

The Pampered Prisoner has more significant choices than the Man in the Pit, and has more autonomy, but the Pampered Prisoner does not have *much* autonomy. The reason is that no other people are able to observe and evaluate his choices, and praise or condemn the Pampered Prisoner for making them.[2] Indeed, the Pampered Prisoner has little basis for choosing between, say, developing an expertise in botany or mastering pole vaulting, or breaking candlesticks. In his solitary world any of these activities derive value only from their intrinsic pleasure. The Pampered Prisoner does not have the choice of rejecting or embracing the values of his community. Without this source of value, options cannot be valued except for their immediate biological pleasure, like deciding when to scratch one's ear. Such a life is not an autonomous life.

One might respond that this argument overstates the importance of the social construction of options. If the Pampered Prisoner devotes his life to discovering mathematical truths, is this not the act of an autonomous person? But suppose he chooses to develop a great musical talent, or to produce sculptures, or to write novels. Do these activities have meaning if there can be no audience for them? These are difficult questions. But they should not obscure the point that many options necessary to an autonomous life derive their significance from their positive or negative effects on other people, and have no value in isolation. The significance of this point will become clearer subsequently.

The Hounded Villager, like the Hounded Woman, formally has a variety of meaningful choices, but they are all dominated—this time, not by the biological necessity of avoiding death, but by the social necessity of avoiding ostracism or, as it is aptly called, "social death." Unlike the Pampered Prisoner, the Hounded Villager has the opportunity to conform to or violate the expectations of others; but this opportunity is merely formal, not real, because the penalty for violating the expectations of others is so severe.

To understand how the state might contribute to or prevent the types of autonomy loss illustrated by the examples of the Pampered Prisoner and the Hounded Villager, it is helpful to redescribe them in terms of the signaling model. The Hounded Villager is someone who lives in a community where people depend strongly on the willingness of others to cooperate with them. Exit is impossible; independent opportunities to obtain gains are nonexistent. Choice of job, spouse, religion, or education are all important signals, and because everyone fears being seen as a bad type, everyone pools around the appropriate signal. The fear of ostracism prevents people from satisfying their intrinsic, non-reputational preferences.

The Pampered Prisoner is a more complex character. The problem here is not conformity. Because no other people exist, signaling is unnecessary. The Pampered Prisoner will discover his intrinsic preferences and satisfy them, unaffected by the opinions of others. An economist might say that the Pampered Prisoner is in an enviable position, but this claim is not plausible. We do not envy the Pampered Prisoner, and the reason is that much of our sense of accomplishment and well-being comes from our considered approval or rejection of the values to which others expect us to conform, and from our consistent action with these judgments despite the pressures put on us by others.

If this is so, then social pressure is an important source of autonomy and a barrier to it as well. If one feels too little pressure to conform to signals and is free to satisfy intrinsic preferences, one lacks autonomy, and if one feels too much pressure to conform to signals, one lacks autonomy.

These ideas can be made clearer by introducing the ideas of "social influence" and "state influence." By social influence, I mean the power of community expectations to limit choices. When everyone must send a particular signal in order to avoid ostracism or to obtain social rewards, the range of choice with respect to important aspects of life may be significantly curtailed. For example, membership in a particular church may become a signal of cooperativeness; then those people who cannot obtain religious fulfillment in that church must either forgo a religiously fulfilling life or risk ostracism by entering a different church.

By state influence, I mean the power of the state to cause conformity even without legally compelling it. I take the following example from Lessig (1995). In the Soviet Union many years ago, motorcyclists wore helmets to protect themselves in case of accident. However, the authorities decided at one point that no one should wear motorcycle helmets, because the helmets were all made in the West. A law was not passed; instead, the state-controlled media announced that anyone who wears a motorcycle helmet is not a patriotic socialist. The response was that people stopped wearing helmets in great numbers. Later, when the Soviet Union began to manufacture its own helmets, it announced that loyal communists wear motorcycle helmets, and people responded by buying and wearing motorcycle helmets in great numbers. This example is trivial but striking because it is clear that people who avoid Western-made motorcycle helmets for fear of being branded a traitor will also avoid books, jobs, political stances, and hobbies about which the state has expressed disapproval, or even which they *believe,* correctly or not, the state might disapprove. When people care about being known as loyal, because the consequence of a reputation for disloyalty is severe punishment by the state, they will signal their patriotism by following even non-binding "suggestions" offered by the state. Despite the absence of a direct legal prohibition of the conduct, autonomy is violated.

It is important to recognize, however, that social influence and even state influence are *constitutive* of autonomy. When social influence and state influence are minimal—say, in a big city where no one knows anyone else—people lack important options: they are Pampered Prisoners. If no one goes to church, then the option of joining a church does not exist. To take another example, the institution of marriage developed gradually from more fluid social practices in which relationships took a variety of forms, all of them less stable than the modern version. Until social influence (with help from church and state) exerted pressure to conform to a single model, in which commitment could be observed and enforced, many people were deprived of a significant option—the option to enter a lifelong monogamous relationship. Many options that are important in life involve participation in communal activities, or require community enforcement; but these communal systems persist only because of social (and maybe state) influence, that is, conformity out of a desire to signal one's cooperative type.

These remarks suggest a way in which the state can increase the autonomy of citizens. Suppose there is an equilibrium in which everyone sends a signal in order to avoid being ostracized. It is possible that people would have more autonomy if they could agree not to send the signal, because the action that constitutes the signal dominates all other options, just as flight from the beast dominates the Hounded Woman's other options. But although this agreement is impossible, the state could subvert the equilibrium by increasing the cost of sending the signal. This can be done in two ways: by taxing or penalizing the action that constitutes the signal, or by decreasing the cost of being ostracized for failing to send the signal.

First, consider a welfare law. A welfare law might allow the Hounded Villager to escape the tyranny of the village's expectations. If she violates these expectations and is ostracized, maybe no one in the village will employ her; but she will be able to live a reasonably comfortable life as a result of welfare payments. Thus, she will not fear ostracism quite as much when deciding whom to marry; and if enough people similarly lose their fear of ostracism, then the sanction will disappear. People simultaneously making decisions in violation of the village's expectations eliminate these expectations' force, just because ostracism requires the cooperation of everyone or almost everyone. Everyone obtains autonomy, and what is remarkable is that *this result can occur without anyone ever actually receiving welfare benefits.*

The welfare law, however, may also destroy autonomy. Suppose that prior to enactment of the law, people engage in important voluntary public functions to signal their devotion to the community. The expectation that one will contribute to the community is not strong enough to cause everyone to contribute, but it is strong enough that those who do contribute feel that they have made an important choice. The welfare law is enacted, and one result is to make people feel more secure. If they lose a job, they do not depend on the community for help. As a result, some of the people who contribute to the community stop contributing; they believe that their efforts are not necessary. If enough people feel this way, contribution to the community will not be seen as an important choice. This important choice being gone, some people have less autonomous lives.

Second, consider a law against discrimination. It is often the case that people in a community discriminate against minorities because this discrimination has developed as a signal of one's loyalty to the community. A law that restricts discrimination obviously enhances the autonomy of the minorities, but it also may enhance the autonomy of the people in the majority. Although it eliminates the option to discriminate, it also creates the option to have important business and personal relationships with members of the minority. Like a law that kills the carnivorous beast, it eliminates one set of options (the Hounded Woman no longer has the choice between running or being eaten), but it creates many new and important options (because the Hounded Woman's choices are no longer all dominated by the need to escape the beast, her choices become meaningful).

The law, however, can also infringe on the autonomy of the majority. If it goes too far, and it succeeds, it prevents the members of the majority from living up to ideals that they find important. If, for example, a law prevents a religious majority from avoiding those people whom they consider heterodox, it may destroy the meaning that the religion gives to people's lives. Religions frequently give people a sense of purpose, but a sense of purpose established by the religion's goal of converting the heterodox cannot survive successful legal restriction. This may or may not be regrettable, but it should be recognized as a violation of autonomy.[3]

This last point raises the following question. If everyone in the status quo is a Pampered Prisoner, can the state encourage autonomy? The answer is that it can, and the analysis is just the same as the analysis above, but converse. The state might be able to create an equilibrium in which a certain kind of act (for example, showing respect for the national flag) has a meaning, whereas earlier it did not (no one did this). The ability to show respect for the flag, to demonstrate one's patriotism, might be a choice which contributes to the autonomy of some people's lives. The state creates this equilibrium by lowering the cost of the signal (for example, inventing the flag). This technique and others can create the conditions under which a significant form of life becomes available: the politically engaged life.

But state influence can be too strong. For example, a state that expects political engagement, and punishes people who are not politically engaged, may produce a stifling conformism inconsistent with autonomy. Are there ways of reducing this danger? At first sight, one might think that a solution would be for the state to refrain from giving "advice." But this is not a satisfactory answer. Government officials must say things about issues of the day. In the United States, government officials constantly criticize drug use, alcohol use, violent movies and television shows, violent lyrics in popular music, careless driving, neglect of children, neglect of religion, and on and on. In a politically free country, it would be perverse to forbid officials to say these sorts of things.

But notice that in the United States no one responds to this "advice" with the alacrity with which the Soviet citizens responded to the "advice" about wearing motorcycle helmets. One reason for this difference is the different attitudes in the two countries to the rule of law. If people know that they cannot be punished for being "unpatriotic," but only for violating certain specified rules, they will not feel the need to respond to government announcements. The rule of law is an important part of the constitutions of the United States and other countries, but there has always been something puzzling about it. Isn't it more important to avoid bad laws than to make sure that all laws, good or bad, are established as "rules" in advance of the conduct to be regulated? The answer to this question is that even very bad laws do not interfere with autonomy as much as a general requirement that people be patriotic, or have good character, followed with vague pieces of advice about what this means. The advice has a disproportionate effect on behavior because people fear being called unpatriotic. Bad rules leave space for autonomous conduct. (There are obviously limitations to this argument, as illustrated by a bad law that says everyone goes to jail.)

Community

In recent years academic attention, especially among political and legal theorists, has focused on the question of "community." There are two causes for this turn of events. First, many people believe that community has declined in the United States, and that this decline has caused a great many social pathologies, from homelessness to the poor treatment of children. Second, many scholars believe that modern liberal political theory does not adequately account for the existence of community and its importance to the well-being of individuals.

I will focus on the first claim, which can be addressed through two questions: (a) What has caused the decline of community, if in fact such a decline has occurred? and (b) How has the decline of community resulted in the various social pathologies, if in fact it has?[4]

What Is Community?

Chapter 2 proposed a provisional definition of community. A community consists of a group of people, most of whom (1) have solidarity, and (2) have enjoyed relationships with each other that have substantial temporal continuity extending into the past and are expected to continue into the future. Solidarity characterizes a group in which individuals have trust relationships, that is, relationships in which people cooperate for the purpose of obtaining some public good and deter each other from cheating most of the time. Solidarity can arise even between strangers who are thrown together for a brief period of time, like passengers in a lifeboat. What distinguishes such a group from a community is that people in a community share a common past and expect a lengthy common future. A common past supplies the focal points around which people signal their types. Newcomers enter the community without changing its character because they enter cooperative relationships only by conforming to the established signals. The lengthy common future matters, because the fear of being deprived of future cooperation deters people from cheating in relationships. To summarize: "community" refers to a group of people who engage in cooperative relationships with each other, and who signal their type to each other by taking actions whose salience results from common pasts, interests, or understandings.

Although communitarians usually stress the involuntary aspects of community, people choose among communities all the time. A person may choose to work in a small firm rather than a large firm because the first has more community. A person may choose to live in a small town rather than a big city because the small town has more community. In both cases, the person realizes that community confers various advantages. In the small town, people are more willing to help members out—to watch over their houses when they are away, or to keep an eye on their children when they play outside. In the small company, other employees may be more likely to make friends with you, do favors for you, and so on. In both cases the more communal environment supplies valuable collective goods while at the same time demanding investment in cooperation. So people who do not value the collective goods offered by the community will choose the anonymous corporation or neighborhood in order to avoid the costs of cooperation (or ostracism). These people will be disproportionately rich. They rely less on communal provision of security, because they can pay for alarm systems and guards, and because they want to avoid the high opportunity costs they incur when required to participate in the creation of public goods.

Nostalgia for community results because when people think about the community into which they were born, that community seems especially attractive compared to alternatives. People who are born into communities and who have lived in them for a long time have preferences that are aligned in such a way that they particularly value the public goods supplied by that community. But when choosing a community as an adult, they must confront the lumpiness of the market—the available communities supply different collective goods from the ones they value the most. There is always a perceived decline in community, because people cannot (usually) enter the community into which they were born, and must choose among the imperfect communities that are most suitable for their current needs as adults. I will return to this point below.

The involuntary aspects of community are twofold. First, the collective goods produced by the community—or what I will call "communal goods"—

depend on the existence of coercive mechanisms of nonlegal enforcement and costly exit. If people can cheat without consequence, either because no one punishes them or because they can leave cheaply if people threaten to punish them, then communal goods cannot be produced. Anticipating this problem, people frequently sink a lot of time and effort into the community. If one buys a big house and fixes it up and invests in relationships, the cost of exit rises. One becomes more trustworthy and a more desirable cooperative partner but only because one has committed one's self to great losses if one is ostracized.

Second, children born into a community are likely to adapt to its values, and thus to be vulnerable to exploitation by members of the community because it is costly for them to move somewhere else once they become adults. Communitarians celebrate the way that communities give children their conception of good, while liberals object to the community's influence over the development of these children, but each side sees only half of the story. One's absorption of a community's values gives one advantages and disadvantages. To be clear about them, assume that a child is born into a community where everyone belongs to religion X and a great deal of social life revolves around the relevant churches. The child grows up with a strong feeling that he can live a satisfying life only if he lives in this town where everyone belongs to religion X. At the same time, he has many skills and tastes that cannot be satisfied in this town, and so is tempted to move to a big city. Compared to a stranger who moves into the town, this person enjoys many advantages. People believe that because he has such a strong commitment to town life, he is unlikely to move away and therefore he is susceptible to sanctions should he breach someone's trust. As a result, people will trust him, and this benefits the person in question, because he can enjoy the gains from cooperative relationships.

The disadvantage of community life is that social norms make people suffer. A bad type who shares the community's values will suffer from discrimination, while his sense that he can live a satisfying life only in the town will prevent him from escaping people's discrimination. A good type with idiosyncratic tastes and values will either suppress these tastes (like the closeted homosexual) or satisfy them and be shunned. Good and bad types will suffer when they, for whatever reason, become a focal point for other people's discriminatory signals, as has often happened to minorities who live in small towns.[5]

How Do Laws Affect Community?

There are two general ways in which the state influences communities. The first is to use taxes, subsidies, and similar devices to raise or lower the value of the collective good that the community produces. The second approach is to refrain from intervening in disputes among members of the community unless the stakes are sufficiently high—in other words, to refrain from interfering with the internal governance of the community.

Following the first approach, governments provide subsidies, tax breaks, and legal privileges to religious organizations, charitable groups of various sorts, even neighborhood associations. The standard rationale is that these groups produce collective goods for their members more efficiently than the government could. Governments effectively delegate some of their functions to the group.

As an aside, one problem with this approach is that the group might distribute the subsidy to the members rather than manufacturing more of the collective good or other benefits desired by the government. For example, if the government gave a church a sum of money because the church provides benefits to the homeless, it is possible that the church would simply reduce the tithe it charges its members, while keeping the poor relief at the initial level. But this will not necessarily happen, especially because the government might respond, if it discovers this behavior, by cutting off the subsidy.[6]

The second approach is more interesting, and is more relevant to our concern about how nonlegal enforcement of social norms maintains or undermines desirable forms of order. This approach is vividly illustrated by disputes involving members of religious organizations.

There are two kinds of dispute. The first is one-versus-many: the attempt to ostracize someone who violates group norms. For example, *Guinn v. Church of Christ of Collinsville*[7] involved a dispute between a church and a female parishioner who violated a church rule by having sexual relations with a man to whom she was not married. The church elders expelled her from the church, announced to the congregation that she was a fornicator, instructed members of the congregation to shun her, and conveyed information about her offenses to other local Church of Christ congregations, where they were publicized. The parishioner won $390,000 in actual and punitive damages for the torts of outrage and invasion of privacy. A divided Supreme Court of Oklahoma reversed, holding that the parishioner was entitled only to damages attributable to punishment that occurred after the parishioner withdrew from the congregation; she would have no remedy for the losses she incurred as a result of the punishment prior to withdrawal.

The result has some appeal. The court approves ostracism from a congregation but not from a town or community. Such an approach circumscribes the power of nonlegal sanctions, allowing people who seek discipline by entering a group to escape it by exiting the group. It reflects an uneasiness about the unconstrained power of social norms and the private institutions that arise to enforce them. This is why courts will sometimes interfere with religious sanctions, even if they rarely disregard them completely.[8] But from the purely ex ante perspective of traditional law and economics, there is no reason for a court to interfere with forms of cooperation that depend on strong sanctions—including the transmission of negative information about a person to people outside the group. The court's interference can be justified only by the theory that such behavior is undesirable, just as state-sponsored shaming punishments are undesirable and the charivari is undesirable. Such a theory must rely on a model that shows that nonlegal regulation may be pathological.

The second kind of dispute is the schism. A majority of a local congregation votes to separate from the church hierarchy to which it belongs. A minority resists, and petitions the hierarchy, which holds that the minority constitutes the "true church," and therefore is entitled to the church property, which is allegedly held in trust by the hierarchy.[9] The court must decide who gets the church property—the building, the fixtures, and so on. It is a sticky business. If the court uses standard contract principles, then it must interpret the "contract" between the local congregation and the hierarchy, a contract which may be a theological instrument. Yet it cannot refuse to intervene when two groups of people claim a property interest in the same real estate, unless it plans to let them resolve the dispute by force. A court might defer to the hierarchy, but that allows the hierarchy to break its own rules. In any case, it is clear that state intervention is necessary.

The cases reflect a tension between deference to the possibly superior nonlegal enforcement mechanisms of religious organizations, and intervention in order to prevent injustice or abuse. A crude utilitarian calculus might tell us that because abuse is rare (for if it were not, people would leave their churches or move to churches where abuses were rare), systematic judicial intervention will not produce much good while interfering with desirable communities and their production of valuable collective benefits. And such a calculus has special weight when (as is usually if not always the case) churches provide fair warning. They state in their charter, or otherwise make clear to their members, that they are committed to resolving all disputes by themselves and without relying on the intervention of the state. Only when the magnitude of the discord overwhelms the church's capacity to resolve disputes should the state become involved, and then only in a careful, deferential manner (E. Posner 1996c).

But this crude utilitarian approach does not seem to be right. Courts' reluctance to entangle themselves in religious disputes is more directly explained by Constitutional constraints. Courts are not as deferential when disputes arise among other kinds of close-knit groups, other than families, and even here courts are not as deferential today as they have been in the past. In the religion cases, the discomfort courts feel about exercising deference is palpable, and not surprising. Deference means tolerating what seems to be an act of injustice in the hope of promoting forms of nonlegal social control of speculative and perhaps dubious value.

One might dismiss the courts' discomfort as the result of a kind of hindsight bias. When evaluating a dispute among members of a religious congregation or any other close-knit group, a court will form its own opinion about the merits of the competing claims. If it agrees with the decision of the group, it need not intervene. If it disagrees, it may have misunderstood events that the group decision-makers more exactly apprehended, or it may have misunderstood the values of that group. If this is the case, courts should exercise self-discipline and enforce nonlegal sanctions, just as they enforce two-party contracts that they do not fully understand and which seem, with the benefit of hindsight, to have been unfair. But courts' discomfort with nonlegal enforcement has deeper roots.

In past centuries in England and the United States, the community was deeply involved in the criminal justice system. Dueling was generally tolerated. Citizens helped capture criminals, judged them (as members of the jury), and participated in their punishment, as I showed in Chapter 6. Today, the police have taken over the first function; the power of juries is circumscribed; and punishment is performed by officials, out of sight of other citizens. Indeed, criminal records are sometimes expunged.

In the past, citizens involved themselves in the regulation of family behavior. The charivari and similar spontaneous gatherings punished people who beat their spouses, committed adultery, and violated other norms of family life. These violators would be mocked, humiliated, and shunned. Charivari has vanished. The regulation of family behavior has been taken over by the agencies of the state.

Although local governments have always distributed poor relief in England and the United States, until recently they did so in partnership with the communities in which the poor resided. Private relief agencies deliberately stigmatized recipients of charity. As government efforts increased, the government also used stigmatization as a weapon for deterring the able-bodied from seeking welfare. But after reaching its apogee in the English workhouse system, the use of stigma in welfare programs has declined. Today, many poor relief programs minimize the stigma attached to those who receive benefits. Of course, when welfare is generous and non-stigmatizing, it may undermine the incentive to work, among people at the margin. And it may crowd out effective forms of private relief, like the informal mutual insurance that was provided by American fraternal societies in the 1920s to as many as one-third of all adults (Beito 1990). But these costs may have been justified by the elimination of the stigmatization of millions of people as an instrument of social policy.

When central authority was weak, trade was dominated by extended families and by ethnic enclaves in larger political entities. Because families provided protection to their members, the threat of exclusion was an effective means of discipline. Similarly, members of ethnic groups who broke promises and violated trade customs risked ostracism, which would throw them at the mercy of resentful native populations. But family and ethnic connections have limits. Quasi-legal institutions such as tournaments and dueling could maintain a semblance of order across a larger fraction of the population, but they had many defects. In the absence of effective mechanisms for resolving disputes, trust mattered. People distinguished those who could be trusted from those who could not be trusted by inquiring about their reputations and observing signals such as dress, gift-giving, manners, and conspicuous consumption. But signals are often ambiguous since they constantly shift in response to changes in costs and beliefs. In societies where honor codes fill the gap left by weak governments, these phenomena are in stark relief. Gift-giving has special significance; dueling is common; public denunciations are frequent and effective; insults are ritualized and public; self-censorship is rife.[10] Behavior that seems trivial to us is invested with fantastic significance. Order of a sort prevails, yet people feel trapped and unhappy, and wish things were otherwise. They admit their abhorrence of dueling in their private diaries, then submit to a challenge and either slaughter or die.[11] So when states finally gain a sufficient amount of wealth and power, they replace the forms of nonlegal regulation with more purely legal institutions, and eventually succeed because the legal institutions are better.

Efforts by states throughout history to domesticate and institutionalize the crowd were often the consequence of a weak government doing the best it could with the materials at hand. This meant using the law to guide existing forms of nonlegal enforcement in desirable directions rather than replacing them with purely legal institutions of enforcement. But the unsatisfactory operation of these institutions was an acknowledged problem, and their gradual replacement with more purely legal institutions was seen as the solution.

To be sure, nonlegal enforcement mechanisms remain strong and pervasive today—as shown throughout this book. Poverty and criminal conviction still produce powerful feelings of shame. Individuals still donate to charities. Religion is booming. The sovereigns of commercial life are custom and reputation, not the law. But the general trend over hundreds of years has been away from nonlegal mechanisms of enforcement, and away from partnerships between law and social norms, and toward the rule of pure law. The trend has been away from incorporation of nonlegal sanctions and toward their suppression. Where pockets of powerful nonlegal enforcement remain, people feel uneasy.

The Decline of Community

If the rise of law has meant the decline of community, and law has replaced dysfunctional forms of regulation, then the decline of community must be thought a desirable trend. This opinion is at odds with an important literature, and so a brief examination of this literature is in order. This literature contains two arguments. The first is that the *market*—or, capitalism—undermines community. The second is that *law* undermines community. Each argument provides different reasons for why the decline of community has been a matter of regret, and should be resisted.

The first argument is associated with thinkers as diverse as Marx, Weber, Schumpeter, Bell, and Hirsch, and has been taken up by modern-day communitarians like Sandel and Taylor.[12] Its most extreme form asserts that capitalism undermines itself by causing the disintegration of the moral rules, left over from pre-capitalist societies, that uphold it. Because the market rewards competitive activity rather than cooperative activity, people abandon age-old beliefs that support cooperation. Yet cooperation is as essential to the market as competition is, and as the ability to cooperate declines, the market will disintegrate.[13]

Here is an example. When I buy something from your store, I might not trust your representations about the widget you sell me and so I may demand a contract. But I at least trust you not to kidnap me and hold me for ransom. But you know that other stores are competing with you, and if they make some kind of innovation, you will be out of business in a second. To survive, it may not be a good strategy to kidnap customers; but, on the other hand, if kidnapping your customers (or, say, every 100th customer) would be a successful competitive strategy, then you, as a rational actor, would do it. Indeed, so would all your competitors. Now, I know this, so I trust you even less, and so I do not go to your store in the first place. As long as pre-capitalist sentiments of trust prevail, I can go to your store and our transaction goes through; but as the logic of the market eats away at these sentiments, business becomes difficult to do.

The problem with the argument is that the market rewards cooperative behavior *and* competitive behavior. People who cooperate can obtain economies of scale unavailable to people who compete. If a particular industry is highly competitive, with the result that agents cannot cooperate with each other even when doing so would be profitable, there is a reward to any entrant who can buy up the agents or start a trade association in which cooperative arrangements are enforced. The more competitive the industry, the greater will be the reward to the first person who can establish cooperative relations. The owner who does not kidnap his customers will attract more business than the owners who do. The argument that the market undermines cooperation by rewarding competition is no more or less plausible than the argument that the market undermines *competition* by rewarding *cooperation.* (Hirschman 1992, p. 139; Conway 1996.)

Moreover, people who are able to cooperate obtain rewards when the market fails. If insurance companies fail to provide insurance against unemployment because of problems of adverse selection and moral hazard, individuals have an incentive to form private insurance groups like the fraternal societies previously mentioned. Religious organizations that provide aid for their members will attract adherents, and large companies that are stable against market shocks and offer protection against layoffs in return for wage concessions will attract employees. Even in the modern world, when the market fails and the government does not intervene, we find robust forms of community.

The second decline-in-community argument connects the decline of community to the rise of law. The argument is that as the state supplies more and more collective goods, people have greater incentives to cheat on community efforts to supply collective goods. But it can be shown that people can be made worse off by this. Government programs may be effective enough to cause people to defect from communal provision of collective goods, while nevertheless being less efficient in producing these same collective goods. Bureaucracies may produce some surplus, but deficiencies in information and incentives result in the production of smaller surpluses than those that would be provided by the communities they undermine (M. Taylor 1976).

The most important culprit is the welfare state, which undermines the community's importance for supplying mutual insurance to its members (Hechter 1987, p. 48). But similar arguments can be made about public education, regulation of commerce, even the police. Thus, the anarchist program calls for the abolition of government on efficiency grounds. Others take a less extreme view, arguing that some but not all modern government-supplied public goods are more efficiently supplied by communities.

The proper response to these arguments is to admit that there are both benefits and costs to legal regulation. Legal regulation, done poorly, will produce fewer gains than communal regulation. The benefit of legal intervention depends on the extent to which it improves conditions prevailing in the absence of legal regulation. But, as we have seen, those conditions can be quite undesirable, and in ways that are not mentioned by those who deplore the decline in community. One must avoid making highly abstract claims about the desirability or undesirability of the decline in community, and evaluate different laws on their own merits.[14]

But I think one can say something even at this level of generality. The decline of community has been caused by, or at least associated with, the rise of two important values: the rule of law, and privacy. The rule of law, as discussed above, can be understood as the appropriate legal response to the dysfunctions of nonlegal enforcement mechanisms. Social norms are hard to describe; they are fuzzy; they drift. People enforce social norms inconsistently, sporadically, unpredictably. Social norms keep a rudimentary sort of order, and are surely superior to chaos, but they provoke a longing for predictability, a longing that can be satisfied only by a wealthy and powerful government. So if a side effect of the rule of law is the loss of certain collective benefits that can be obtained only through nonlegal enforcement, that might seem a straightforward improvement.

Privacy is not a theme I have stressed in this book, although its connection to the subject might seem obvious. To say that a person values privacy is the same thing as saying that he fears the enforcement of social norms. The steady increase in the commitment to privacy over the last century is tightly linked to the decline of community and the rise of the rule of law. A person's concern for his privacy is not just a concern about concealing discreditable or embarrassing information about himself (R. Posner 1998b); it is a concern about protecting himself from being shunned because of his failure (currently or in the past) to conform to unpredictable signaling equilibriums. Put another way, the rule of law and the legal protection of privacy are two arms of a general approach to solving the pathologies of nonlegal mechanisms of enforcement. A criminal justice system replaces the crowd with a bureaucratized system, but this system will inevitably be influenced by the crowd in times of tension. When the crowd influences the criminal justice system, enforcement of the rule of law by independent judges ensures that the pathologies of nonlegal enforcement are avoided. In addition, social norms will influence people even when these norms are not directly enforced by the authorities. To minimize the influence of nonlegal enforcement of such norms, the state protects privacy, which deprives the crowd of the information it needs to inflict sanctions.

A related trend is the decreasing protection of reputation under defamation law. The Supreme Court has weakened the common law of defamation by vigorously enforcing the free speech clause of the Constitution. The consequence of this doctrinal innovation is that reputation has become a less valuable asset than it used to be. If people can call me a liar and escape legal retaliation, then I do not have much incentive not to lie; after all, whether I lie or not, the accusations might stick. So in my dealings with others, I have to rely more on contract law when I make promises and rely on the promises of others than I could when defamation law was more substantial.[15] Now, one view of these developments is that the Supreme Court has erred: the subversion of defamation law is the unintended consequence of a simple-minded commitment to the marketplace of the ideas, according to which lively debate has only benefits and no costs. But a subtler position is that the Supreme Court has absorbed a widespread view that nonlegal regulation based on reputation is dysfunctional, so its weakening at the hands of free speech doctrine is not such a bad thing.

NOTES

REFERENCES

ACKNOWLEDGMENTS

INDEX

Notes

1. Introduction

1. New York v. P., 90 A.2d 434 (NY 1982). Courts have begun to back away from this very old view, as the stigma against illegitimacy continues to wane. For example, West Virginia v. Stone, 474 SE.2d 554 (WV 1996).

2. See, for example, Brandt v. Board of Co-Op Educ. Services, 820 F.2d 41 (2d Cir. 1987).

3. The T. J. Hooper, 60 F.2d 737 (2d Cir. 1932).

4. Compare Golding v. Golding, 581 N. Y. S.2d 4 (1992).

5. 387 U.S. 483 (1954).

2. A Model of Cooperation and the Production of Social Norms

1. On these problems and possible solutions relying on traditional rational choice models, see Olson 1965, Hardin 1982, Hechter 1987, Coleman 1990.

2. The conditions under which repetition produces the cooperative outcome in the repeat prisoner's dilemma are discussed in any game theory or economics textbook; see, for example, Kreps 1990b. Relevant discussions can be found in Telser 1980, M. Taylor 1976, and many other places.

3. People signal other characteristics, for example, their cultural competence. When people enter relationships, they seek people who will not misunderstand statements and behaviors because these actions have different meanings in the culture from which they come. People also signal their health, their intelligence, and so on. Such signaling may give rise to norms. But a complete analysis of these phenomena would be overwhelmingly complex.

4. Descriptions of the model can also be found in any game theory textbook; see, for example, Fudenberg and Tirole 1991 and Gibbons 1992. The use of signaling models to

explain cooperation, social norms, and norm-like behavior (like fashion) can be found in Klein and Leffler 1981, Bernheim 1994, and Pesendorfer 1995.

5. The absence of a consensus on an appropriately stringent equilibrium concept muddies the analysis, but this muddiness is unavoidable given the state of the literature. See Kreps 1990c.

6. Superficially direct forms of signaling—like displaying one's bank statement or mutual fund shares—are, on reflection, quite inadequate. Documents can be forged, and can be shown only to a limited number of people. Financial resources can be shifted around—for example, one can borrow lots of money, put it in a bank account, obtain a statement, then repay the money a few days later. So this behavior is really just cheap talk (see Farrell and Rabin 1996), like telling people that you are rich, which people do anyway with less hassle. Credible signaling requires more complex behavior and, as discussed in later chapters, complex institutions supplied by the market.

7. Tattooing has a long and interesting history. It has always been used both voluntarily in order to mark oneself as a member of a group (frequently, religious) or involuntarily to mark a person as a criminal (see Chapter 6). See, for example, Gustafson 1997, C. Jones 1987. We associate scarring with tribal societies, but encounter it as recently as the nineteenth century, when in certain German university towns a "face disfigured by scars was a passport to a good marriage" (Kiernan 1988, p. 201)—dueling scars, of course. On its recent revival, see Austin 1999.

8. When we observe people in a foreign culture or in the past (for example, in a movie) acting inconsistently with our norms (that is, their clothes conform to some long-dead fad), we instinctively react with something like amused sympathy. We don't feel contempt because we know that the whole population cannot consist of bad types; it's as though we pity them for being born at the wrong time, the way we might pity a person who was never taught proper manners as a child. A minor act of imagination allows us to see that we look as ridiculous to others.

9. Elias' (1982) famous description of changes in manners in the Middle Ages also stresses the arbitrariness of manners, but he argues that the development of manners reflected efforts by members of upper classes to distinguish themselves from members of lower classes. I suspect that this conclusion is an artifact of his sources, that within a "lower class" an independent development of manners occurred for the reasons described in the text.

10. I am relying here on the idea of the focal point. See Schelling 1960.

11. That hygiene, for example, is as much a matter of statistical averages as of health, is well-understood. See Elias 1982, p. 158, as discussed in Kagan and Skolnick 1993, p. 81.

12. My example, and many models, cheat a bit, producing the discontinuity in behavior by building in assumptions about discontinuous choices. However, Bernheim (1994) shows that a similar result can be obtained even if the assumptions are relaxed and continuous choices are permitted.

13. Much more can be said about why certain actions become signals. When two actions have the same cost structures, either could serve as a signal, and whether one or both serve as a signal depends on the outcome of an n-person coordination game played among the good types. (The bad types will not send the signal.) Once an equilibrium is achieved, one might call the equilibrium signal a "convention" (Lewis 1969, Ullman-Margalit 1977, Schotter 1981). To explain why over time people will settle on one convention rather than any other, one needs to rely on an evolutionary theory—the basic lesson of which is that conventions that are more robust against random shocks are more likely to prevail over

time than those that are not. See, for example, Young 1998a. As Chapters 3 and 10 discuss, evolutionary game theory has much promise, but at its current stage its implications for the questions I am interested in are limited. See also Boyd and Richerson 1985, and Kraus 1997.

14. The commitment model is from Schelling (1960). In the commitment model, the agent takes an action that reduces his payoff from taking an action different from that which the cooperative partner prefers. In the signaling model, the agent's action does not affect future payoffs, but reveals information about the agent's type. Hardin (1995) uses the commitment model to explain aspects of group behavior and certain kinds of social norms.

15. Sunstein 1996. The idea, of course, has a long history in sociology.

3. Extensions, Objections, and Alternative Theories

1. See Sober and Wilson 1998 for a survey.

2. See Margolis 1982 for an approach to altruism that is more complex than the approach discussed above.

3. See Frank 1985 on status, and Hirsch 1976, whose approach is slightly different from that described in the text, and also Elster 1989, p. 261, on the related concept of envy. See S. Jones 1984 and Bernheim 1994 on conformity.

4. An alternative model of the emotions can be found in Huang and Wu 1994.

5. Treating social norms as exogenous, that is, assuming that they exist and that they influence behavior, but not explaining what causes them to arise and persist, is very common, especially in the legal literature. See, for example, Lessig 1995, Sunstein 1996, R. A. Posner 1997. Cooter (1996, 1998a, 1998b) tries to give this approach more purchase by making minimal assumptions about the internalization mechanism. Bowles (1998) provides a helpful survey of methods for endogenizing preferences, but although he is optimistic, the main lessons of his article are the vastness of the problems that must be confronted, and the crudeness of the methodologies that might deal with them. See also Rabin 1998.

6. Mailath (1998, p. 1372) says that evolutionary game theory models are currently "too stylized to be used in direct applications." Sethi and Somanathan (1996) show ways in which cooperation can prevail in common pools, but again it is difficult to extract concrete normative implications from their analysis.

4. Gifts and Gratuitous Promises

1. An illustration is the uproar created by the retired news anchor David Brinkley's decision to serve as a spokesman for a large agribusiness. "He built his reputation on being this acerbic, no-nonsense guy who would never lie to you. *What is he doing giving his reputation for integrity to ADM for money?*" Kurtz 1998 (quoting Daniel Schorr, emphasis added). Schorr also says that he turned down $1 million to be a spokesman for Avis: "I have spent 50 years building the reputation I have, and the first time I was on the air with this, I would throw away that reputation[.]" *Id.*

2. See especially the work on potlatch, such as Adams 1973. Adams' account of the function of potlatch among the Gitksan could be easily recast in terms of the signaling model. The potlatch is a signal—easily observable, clearly wasteful—which both enhances reciprocal relationships and leads to destructive competitions for status.

Courts have also understood the connection between gift-giving and exchange. See *Hamer v. Sidway,* 124 N.Y. 538, 27 N.E. 256 (1891), in which the court enforced a man's promise to pay his nephew $5000 on his 21st birthday if he gave up smoking, swearing, and gambling; see also *Allegheny College v. National Chautauqua County Bank,* 246 N.Y. 369, 159 N.E. 173 (1927), which involved a charitable contribution.

3. See also Kranton 1996a, 1996b.

4. A subtle example of this phenomenon is corporate donation to charity. In one instance, authorities enacted laws protecting local businesses from takeover threats at the request of managers. The authorities feared that a takeover by outsiders would end a large local business's periodic donations to local charities. See Romano 1987, p. 121.

5. This is not to say that universities can always resist skating as close to the line as possible. Schmitt (1995) describes the method of the University of California, Irvine. "'The potential donor, Mr. Barclay,] asked the magic question: What does it cost to put your name on something?' recalls Terry Jones, a former UCI development officer who handled the Barclay gift . . . Mr. Jones says that Mr. Barclay was pitched several 'naming opportunities'—projects that would be named in his honor in exchange for a contribution—and 'his eyes kind of lit up at the theater.' After a series of negotiations, he agreed to the $1 million, payable in $200,000 installments over five years. After that, the Barclays were treated more or less like royalty. Mrs. Barclay, for example, was 'knighted' at an elaborate ceremony put on by the university's madrigal society. The couple was saluted at luncheons, cocktail parties and the gala review. And they were given other tokens of affection, including his-and-hers windbreakers bearing the university's Anteaters logo."

6. Another possible motive, discussed in the economic literature on "warm-glow" giving, is that people take pleasure from the act of giving itself. See Andreoni 1990.

7. For example, suppose that donor gives a painting to donee. If donor values the painting at $100, donee values the painting at $200, and donor obtains utility from the donee's happiness at, say, a 40 percent discount, then donor obtains $120 worth of utility from donee's acquisition; subtracting the $100 loss, the donor nets $20. The donee, of course, gains $200.

8. One study showed that anonymous gifts accounted for 1.29 percent of the donations to the Pittsburgh Philharmonic, and for less than 1 percent of the donations to Yale Law School, Harvard Law School, and Carnegie Mellon University. See Glazer and Konrad 1996. Even in these cases, the donor's identity is most likely known to the managers of the donee (if not to the wider public), and the donee may be managed by prominent people whom the donor wants to impress, for example, the board of trustees of a university.

Glazer and Konrad also point out that the motive of altruism cannot explain why people typically make gifts at the lowest value in the range by which donors are classified in published reports. For example, 93 percent of those who contributed to the Harvard Law School Fund in the category of $500-$999 made contributions of $500. Id. People motivated by altruism would presumably give gifts along a broader distribution.

9. For evidence that altruism is an insufficient explanation for gift-giving, see Cox 1987. Cox argues that a desire for exchange motivates most gift-giving; for example, parents give gifts to their children in the hope that their children will later support them. This is similar to the trust explanation.

10. Economic theories of status trace their origin to Veblen 1992.

11. Wealthy people who do not give, according to interviews of other wealthy people, are considered "warped," "revolting"; they are "looked upon with disdain, disfavor, and are highly criticized." See Ostrower 1995, pp. 14–16.

12. I ignore two arguments of Titmuss 1971: first, that in certain markets the quality of the product depends heavily on non-observable attributes and that only altruists will self-screen so as to produce products only with those attributes (Titmuss's example is blood); second, that the cultivation of altruistic tendencies produces social goods that are independent of the aggregation of the satisfaction of private preferences.

13. Kaplow points out that a bargain will not occur because an altruist would not gain from a return payment. A gift plus a return payment is equivalent to a smaller gift; but if the original gift would not be made, then neither would the donor accept the bargain. See Kaplow 1995.

14. Further discussions can be found in Stark 1995. The first argument is also made by Buchanan 1975; Shavell 1991.

15. Bosses hate getting gifts from subordinates, yet a survey showed that one-third of employees in the United States planned to buy their bosses Christmas gifts in 1997. Quintanilla 1998.

16. Ostrower notes that wealthy individuals solicit donations from each other for the benefit of the institutions on which they serve as board members. See Ostrower 1995. Thus, what often appear to the outsider to be one-sided gifts from rich people to cash-strapped institutions may in fact reflect just one side of a reciprocal relationship.

17. Ostrower argues that elite philanthropy is mainly a system used by elites to preserve and enhance their status. She correctly points out that altruism cannot be a sufficient explanation of elite philanthropy, because it does not account for the nature and the target of elite gift-giving. However, she does think that altruism plays some role; she points to the practice of allowing the nouveaux riche to "buy" trustee positions (a chief symbol of status; see, for example, id. 38), and argues that the incumbents permit this practice because they care about the charity. Id. 141. If, however, one broadens the perspective, and assumes that status is not intrinsically desirable but matters as a method for distinguishing good types (wealthy and cooperative) from bad types, the reason for allowing newcomers is clear: it is to ensure that the elite class remains the elite, that is, composed of people one would profit from interacting with.

18. Compare Glazer and Konrad 1996. For an example of the way a donee will engage in elaborate techniques for publicizing the gift, see the discussion of the Barclay gift, above.

19. Charitable giving, especially by the wealthy, is skewed toward high-profile gifts to cultural institutions. See Clotfelter 1992; Ostrower 1995. The exception is contribution to religious organizations.

20. The potlach is a superb example of gift-giving that enhances cooperation by serving as a signal but ends up leading to destructive status competitions. On its importance for enhancing cooperation, see, for example, Simeone 1995, p. 165; Kan 1989, ch. 9; more information about the attempts of missionaries to prohibit it can be found in Simeone 1995, ch. 2. See also Johnsen 1986.

21. See Trebilcock 1993, pp. 170–87; Goetz and Scott 1980; R. Posner 1977; Shavell 1991; Eisenberg 1979; Farnsworth 1995; Gordley 1991; Kull 1992.

22. This point is emphasized by Eisenberg 1979, p. 17, and Kull 1992, p. 50. Eisenberg argues that transfers are harder than promises for a putative donee to fake, and that transfers are more likely than promises to have been a subject of deliberation on the part of the donor.

23. This question has consumed the attention of many commentators; see, for example, Eisenberg 1979.

24. See *Scholes v. Lehmann,* 56 F.3d 750 (7th Cir. 1995). A similar case is *Young v. Crystal Evangelical Free Church,* 82 F.3d 1407 (8th Cir. 1996), where the court held that donations (based on tithing) to a church were (constructive) fraudulent conveyances. The court pointed out that while the donors may have received benefits from the church, they would have gotten them whether or not they donated. There was no quid pro quo (motive was something akin to altruism). (The court held that application of fraudulent conveyance law violated the Religious Freedom Restoration Act.)

25. This was apparently the purpose of the original fraudulent conveyance law, the Statute of 13 Elizabeth.

26. Recently, Congress enacted a law that exempts certain charitable organizations from fraudulent conveyance attacks in bankruptcy. See 11 U.S.C. §548(a)(2).

5. Family Law and Social Norms

1. See Scott and Scott 1998. Here and elsewhere I rely on their setting up of the problem, although my conclusions differ. See also Silbaugh 1998, pp. 100–08.

2. Bishop's (1984, p. 250) similar model has parties signaling their interest in maintaining an exclusive relationship with their partners.

3. Different legal and normative regimes solve this tension in different ways. One way is to forbid premarital sex but allow annulment of marriages when one partner is not sexually healthy. The engagement then is celibate. Another way is to allow sex during engagement or an engagement period, which can be broken only if one partner is not sexually healthy. Typically, the engagement period will be preceded by a pre-engagement courtship when parties are celibate. A third way is to allow pre-marital sexual activity short of intercourse ("bundling" may have been an example of this). One finds all three patterns in various times and cultures.

4. See also Lundberg and Pollak 1993, pp. 993–95. A very useful treatment of bargaining between husband and wife, which takes account of social norms as well, can be found in Agarwal 1997.

5. The idea that spouses may take on certain roles that enable them to monitor each other effectively, or, more generally, that spouses will resolve disputes along general principles that are set up in advance, has interesting parallels with Kreps' theory of corporate culture. See Kreps 1990a.

6. A historical example can be found in Weber 1976.

7. They would also punish people who violated norms at the courtship stage, including (1) rules against interference with other people's relationships during courtship; (2) norms governing the giving of gifts, parties, dances, and pre-marital sex; and (3) the stigma that attaches to the Don Juans, the people who mimic the good types in the hope of fooling people into believing they are serious about a relationship. For brevity, I ignore these complications.

8. See Davis 1975, pp. 97–123; Wyatt-Brown 1982, pp. 435–61; Ingram 1984; Thompson 1993, pp. 467–538.

9. Bishop makes a similar suggestion. See Bishop 1984, pp. 252–54.

10. See the similar argument in Allen 1990. Marriage reform by ecclesiastical and civil authorities was often a response to the chaos that existed in the absence of requirements that marriages be licensed and weddings be public—the endless disputes about whether or not one person married another; for examples and discussion, see Helmholz 1974. Author-

ities frequently punished marital opportunism, such as adultery, so long as it was sufficiently openly practiced that problems of proof could be overcome. See Harrington 1995, pp. 249–50.

11. For historical evidence, see Levine and Wrightson 1980.

12. A similar point is made in Trebilcock and Keshvani 1991, p. 558. See also Lundberg and Pollak 1993.

13. For historical evidence of efforts by parents to control the marital choices of their children, see Harrington 1995.

14. This is also true if the costs of entering marriage are too high; for an example from nineteenth century Germany, where wealth and citizenship requirements for marriage resulted in a great deal of cohabitation and illegitimacy, see Abrams 1993.

15. See Chapter 9.

16. However, the stigma of illegitimacy continues to play a role in judicial opinions. See Chapter 1.

17. See Bix 1998, pp. 158–62. An interesting general discussion of the meaning of marriage, and how legal policies change that meaning, can be found in Silbaugh, 1998, pp. 81–83, 108–17.

6. Status, Stigma, and the Criminal Law

1. One important reason why gangs provide gratuitous protection to a community is that they are highly vulnerable to arrest if community members choose to inform on them. When discipline breaks down, and the gangs inflict too much harm on the community without providing enough protection, members of the community will call in the police. See Jankowski 1991, pp. 203–06.

2. This point is familiar; see, for example, McAdams 1996, pp. 2281–82.

3. Gangs are sometimes popular in their communities because they can provide more effective protection than police can. There are several reasons for this. First, they have better information about criminal activities. Second, they do not have to worry about civil rights: they can question or tail anyone they want to. Third, they can dispense harsher punishments. See Jankowski 1991, pp. 180–93, 260.

4. For evidence that reputational harm is not predictable, see Lott 1992. He shows that "the dollar amount taken in embezzlements or frauds appears to have both a very insignificant and economically very small impact [on the reduction of income after conviction], while it is actually positive and significant for larceny and theft, though the coefficient is quite small." (p. 598) The seriousness of the crime is, for all intents and purposes, unrelated to the size of the reputational sanction. See also Waldfogel 1994a, 1994b.

5. For a discussion of the possibility that shaming penalties should be imposed on white collar criminals, see Kahan and Posner 1999.

6. This can be explained with the commitment model; see Hardin 1995.

7. For example, Sacco and Vanzetti, the Rosenbergs, Hiss. See Russell 1986, pp. 206–12, for a discussion of this phenomenon in the Sacco and Vanzetti cases, and in other cases as well.

8. Jankowski 1991, p. 272, says that former prisoners obtain status among other gang members.

9. See Jankowski 1991, p. 81: "To the Los Angeles gangs, taking drugs is a way of separating themselves from others in the community, a way to make a statement that they are

not going to follow the rules of a society that asks them to conform, but offers them only inferior jobs in return."

10. See also Robinson and Killen 1997 on warning labels.

11. See Diehm 1992 for further examples.

12. See Lott 1992, Waldfogel 1994a and 1994b, on the reputational effect of criminal records.

13. Such self-abasement is an important part of law enforcement strategy, especially during the parole period. See Jankowski 1991, pp. 276–80.

14. Compare Braithwaite's (1989) argument about the importance of "reintegrative shaming." See also Bushway (1997).

15. A similar point is made by Rasmusen 1996, who uses a related model.

7. Voting, Political Participation, and Symbolic Behavior

1. There is, however, a vast antropological and historical literature. See, for example, Wilentz 1985.

2. Certain equilibrium refinements allow us to narrow the range of equilibria, but these refinements themselves are not particularly plausible, in effect transferring the uncertainty from the prediction to the methodology.

3. For a different analysis of flag desecration issues, see Rasmusen 1998.

4. When the relevant community is not the nation, but a smaller community, such as a university or small town, a similar result obtains. See Loury 1994.

5. See Knack 1992, p. 143. In one poll 41 percent of regular voters agreed with the following reason for voting: "My friends and relatives almost always vote and I'd feel uncomfortable telling them I hadn't voted." Id., p. 137.

6. See Lupia and McCubbins 1998. This book suggests that the fact that voters can usually vote for those who serve their interests does not imply that voters incur costs to gather information so that they make correct political decisions, which would be inconsistent with the signaling theory. It shows that a variety of shortcuts enable people to make reasonable political choices while relying on limited information, information which, I would argue, they obtain casually, for other purposes.

7. This is an old theme. For its appearance in fascism, see Mosse 1985.

8. See Gambetta 1988, Przeworski 1998, and other chapters in Elster 1998a.

9. There is an interesting parallel to Stephen Holmes' (1995) discussion of "gag rules," a form of self-censorship which he appears to believe is desirable, or often desirable, because it takes contentious issues, like disagreements over religion, out of public debate, enabling citizens to cooperate on other matters. The discussion in my text may explain how gag rules come into existence—a lacuna in Holmes' account—while not providing much reason to believe that they will be desirable.

10. For a general discussion, see E. Posner 1996b.

8. Racial Discrimination and Nationalism

1. Compare Roback 1989, whose approach is similar, except that she assumes that people have a preference for conformity that extends to choices about association with members of other races. See also McAdams 1995.

2. Also compare Hardin's interesting discussion; see Hardin 1995, pp. 79–91.

3. Discrimination on the basis of a mutable characteristic can, in theory, prevail, when

the victims of discrimination maintain this characteristic as a commitment device. Membership in a disfavored religious organization is not immutable; people often maintain and even publicize their memberships, despite discrimination, in order to enhance their commitment to the group, which may allow them to obtain a larger portion of the group's gains.

4. The politicization of race by political entrepreneurs is discussed in Roback 1989.

5. See Netanyahu 1995, pp. 1052–54. Netanyahu puts more emphasis on the interest of the majority in seizing economic and political power from the Jews (in Spain, the conversos), than on the importance of cooperating on a national level, but he does elsewhere identify the demand for national unity in Spain as an important cause of the persecution of the conversos. Id., p. 1004. Spain united around its opposition to the Jews; when religious Jews vanished, racial Jews had to be invented.

6. The initial problem for leaders in a power vacuum is to persuade people that the expedient enemy (those who have land, factories, or other property that the leaders want) are also an ethnic group, and to do so they will appeal to and manipulate history in order to make their arguments convincing. Once one group becomes persuaded that it is entitled to land belonging to people outside the group, then outsiders will quickly form groups now based on the plausible argument that the first group poses a threat. Thus, nationalism is characterized initially by ferment and confusion; then memberships lock into place and are robust against further shocks. This seems, at a very abstract level, to have been the pattern in Yugoslavia. See Malcolm 1996.

7. As noted by Koppelman (1996), pp. 101–103.

8. Kuran 1998 makes a similar argument.

9. See Kuran 1998, and the many examples in Hobsbawm and Ranger 1983.

9. Contract Law and Commercial Behavior

1. See, for example, Cheung 1973, Ostrom 1990, Ellickson 1991, Benson 1992, Greif 1993, Greif, Milgrom, and Weingast 1994, Landa 1994, E. Posner 1996c.

2. See also Craswell and Calfee 1986 for an analysis of judicial error in tort cases. Williamson (1985) also emphasizes contractual uncertainty in the presence of asset specificity.

3. The argument is illustrated by the following matrix, where it is assumed that the payoff from cooperation is 3, the payoff from mutual defection is 1, the payoff from cheating the other party is 4, and the payoff upon being cheated is 0.

	Cooperate	Defect
Cooperate	3, 3	−C, 4−C
Defect	4−C, −C	1−C, 1−C

If C is, say, 6, then we have:

	Cooperate	Defect
Cooperate	3, 3	−6, −2
Defect	−2, −6	−5, −5

The Nash equilibria are {cooperate, cooperate} and {defect, defect}, but as mentioned in the text the former is more plausible, being Pareto superior to the latter.

4. An alternative theory follows from the signaling model. When a person enters a legally binding contract, he incurs an immediate cost equal to the present value of the stochastic loss resulting from incompetent judicial enforcement sometime in the future. Incurring this cost serves as a signal of a low discount rate, just as the marriage vow does. The commitment theory differs by requiring that the actual payoff, in a future round, be affected by the judicial action in such a way that even a person with a high discount rate would not defect in that round. Thus, the commitment theory is more robust than the signaling theory, and that is why I emphasize it in this chapter.

5. For descriptions, see Bartlett 1986, pp. 103–126; Lea 1967; and Nelson 1890. The negative tone of the descriptions is probably due to their focus on the decline of the institution, when it was dysfunctional, compared to the more modern-seeming procedures that were gradually replacing it. It is hard to believe that trial by battle was completely dysfunctional, however; it may have been superior to anarchy.

6. W. Schwartz et al. 1984; Kiernan (1988, p. 144) describes the duel *au mouchoir,* "with the antagonists close enough for each to hold a corner of a handkerchief . . ."

7. For discussions of legal form, see Kennedy 1973, Ayres and Gertner 1989, Kaplow 1992.

8. See O. Holmes 1963, pp. 227–230; Hand's most famous remark on this subject can be found in *Hotchkiss v. National City Bank of New York,* 200 F. 287 (S.D.N.Y. 1911).

9. This is the point of the literature on this, cited above.

10. 133 N.W.2d 267 (WI 1965).

11. Compare Epstein 1992b, 1992c, and Goetz and Scott 1981, Gillette 1998, Cooter 1994b, Kraus 1997. For a historical perspective, see Thompson 1993, and for a philosophical perspective, see Craswell 1998.

10. Efficiency and Distributive Justice

1. Not necessarily that the former donor had a high discount rate; perhaps just that this person could obtain higher payoffs from a relationship with someone else.

2. The ambiguous welfare implications of signaling equilibriums are discussed in any game theory textbook. Their legal implications have also been recognized. See, for example, Ayres 1991.

3. See Frank 1985, McAdams 1992, pp. 72–76. McAdams (1992, pp. 83–91) argues that people will substitute to more desirable behavior, but it is hard to see why this would be so.

4. Skepticism about the efficiency of small group behavior, relying on a related perspective, can be found in Hardin 1995.

5. Dueling was widely seen as ridiculous shortly before its demise, and was satirized. Kiernan 1998. Clothing fashions seem ridiculous as soon as they pass. Dysfunctional norms often make people feel ridiculous but helpless: for example, wearing ties or high heels, or attending a party for someone everyone hates, or a celebration of someone who has done nothing good, or receiving gifts that one does not want from people who do not want to give them. See Quintanilla 1998 ("no Christmas tradition irks [bosses] more than sleigh-loads of gifts from subordinates").

6. See Sugden 1986 for a numerical example, a fuller discussion, and qualifications.

7. Young (1998a, 1998b) shows that as long as behavioral regularities are subject to persistent shocks, which is a plausible assumption, over a long enough time we can expect more efficient norms to prevail more often than less efficient norms. This is not an argument against government intervention, because if the government has sufficient information, it can change inefficient conventions more rapidly, and perhaps protect efficient conventions from shocks as well.

8. This is a theme of Waxman 1977. Bankruptcy laws follow a parallel history. Earlier bankruptcy laws were designed to enhance the stigma of default. In some countries the bankrupt was stripped naked in public and humiliated in creative ways. See Whitman 1996. The debtor's prison—I do not know whether by intention or not—also stigmatized its inmates. The abolition of debtor's prison in the nineteenth century and subsequent bankruptcy legislation were designed to eliminate the stigma. The American Bankruptcy Reform Act of 1978 went so far as to change the legal designation of the beneficiary of the bankruptcy system from "bankrupt" to "debtor" in order to erase the stigma.

9. See, for rare examples of attention to this problem, Lindbeck, Nyberg, and Weibull 1999, Besley and Coate 1992, Moffitt 1983.

11. Incommensurability, Commodification, and Money

1. For modern discussions, see, for example, Raz 1986, pp. 321–66; E. Anderson 1993; Chang 1997; Sunstein 1994; Nussbaum 1998; Adler 1998. An interesting discussion, unfortunately ignored in the modern literature but full of relevant insights and anticipations, can be found in Simmel 1978.

2. It should be clear that I am not presenting a philosophical argument that is intended to defeat the philosophical argument of the incommensurabilists. My argument is sociological, in the broad sense of the word.

3. It is possible that, instead, markets will segment.

4. Advertisers recognize this dilemma, and sometimes they exploit it by parodying their own efforts to persuade people to buy their products.

5. Compare Elster 1988a, whose description of the riot of human motivations provides a phenomenologically attractive but methodologically sterile basis for understanding social behavior.

6. See Kelman 1981, who criticizes cost-benefit analysis for assuming that environmental goods can be monetized, and for legitimating that practice.

7. Compare the opening scene of *King Lear.* To be sure, Cordelia opted out of Goneril and Regan's bidding war; but look at what happened to her.

8. Compare Simmel's discussion of the alienating effect of money; see Simmel 1978.

9. See the discussion of "avoision" in Katz 1996.

10. I bracket the question of how the government should make tradeoffs, on which see Adler and Posner (1999).

11. See Radin 1987. Although Radin prefers the term "commodification," this term and "incommensurability" are used interchangeably in the literature.

12. As noted before, reputational concerns do not only distort people's claims; they distort people's behavior. This suggests that economists should be skeptical not only of survey data (a skepticism that is frequently expressed in the literature) but also of market behavior when trying to determine preferences.

12. Autonomy, Privacy, and Community

1. So I am relying on a very thin idea of autonomy. Many philosophers insist on a thicker view. Gerald Dworkin (1988) argues that autonomy should be defined in terms of metapreferences. Nussbaum argues that autonomy should be defined in terms of the ability to make specific types of choices, for example, those pertaining to central areas of life. See Nussbaum 1999. See generally Hill 1991, and for an intellectual history of the concept, Schneewind 1998. It hardly needs to be said that this simplification is intended to keep the discussion manageable, and not to be a philosophical argument.

2. There is the further point that the Pampered Prisoner is not able to enjoy meaningful relations with other people, surely a prerequisite to an autonomous life for many people. But this is a separate point, which could be addressed in a more complex example.

3. These examples raise again the possibility that the indeterminacy of the argument results from an inadequate conception of autonomy. Thicker conceptions of autonomy may produce more determinate results; the problem is that these thicker conceptions are themselves controversial, so the indeterminacy is not avoided, but moved to a different location in the argument.

4. Although there are obvious connections, I will not discuss the relationship between the cooperation model and theories of spontaneous order (see especially Hayek 1973; also Buchanan 1977, Benson 1990). This complex subject is best left for future work.

5. The misery and conflict in real communities is well-documented. See, for example, Kniss 1997, on the Mennonites.

6. On crowding out of privately supplied public goods, see Bergstrom, Blume, and Varian 1986.

7. 775 P.2d 766 (Ok. 1989).

8. A rare example is *Bear v. Reformed Mennonite Church,* 341 A.2d 105 (Pa. 1975).

9. See, for example, *Jones v. Wolf,* 443 U.S. 595 (1979).

10. See Greenberg's (1996) evocative description of these behaviors in the old South.

11. A theme of Kiernan 1988; also see his discussion of anti-dueling societies.

12. See, for example, Schumpeter 1950, Bell 1996, Hirsch 1976, Sandel 1982, Putnam 1993, and C. Taylor 1995. A lucid account can be found in Hirschman 1986, ch. 5.

13. Also compare Elster's more measured comments (Elster 1988a, pp. 284ff.), and Hirschman 1992.

14. See the useful discussions in Hardin 1995 and Macedo 1996.

15. I thank John Goldberg for drawing my attention to this line of argument.

References

Abrams, Lynn. 1993. "Concubinage, Cohabitation and the Law: Class and Gender Relations in Nineteenth-Century Germany." *Gender and History,* 5: 81.

Adams, John W. 1973. *The Gitksan Potlatch: Population Flux, Resource Ownership and Reciprocity.* Toronto: Holt, Rinehart and Winston of Canada, Limited.

Adler, Matthew. 1998. "Incommensurability and Cost-Benefit Analysis." *University of Pennsylvania Law Review,* 146: 1371.

Adler, Matthew, and Eric A. Posner. 1999. "Rethinking Cost-Benefit Analysis." *Yale Law Journal,* 109: 167.

Agarwal, Bina. 1997. "'Bargaining' and Gender Relations: Within and Beyond the Household." *Feminist Economics,* 3: 1.

Akerlof, George A. 1984. *An Economic Theorist's Book of Tales.* Cambridge, England: Cambridge University Press.

——— 1997. "Social Distance and Social Decisions." *Econometrica,* 65: 1005.

Allen, Douglas W. 1990. "An Inquiry into the State's Role in Marriage." *Journal of Economic Behavior and Organization,* 13: 171.

Allen, Jeffrey E., and Robert J. Staaf. 1982. "The Nexus Between Usury, 'Time Price,' and Unconscionability in Installment Sales." *U. C. C. Law Journal,* 14: 219.

Anderson, Benedict. 1983. *Imagined Communities: Reflections on the Origin and Spread of Nationalism.* London: Verso.

Anderson, Elizabeth. 1993. *Value in Ethics and Economics.* Cambridge, Mass.: Harvard University Press.

Andreoni, James. 1990. "Impure Altruism and Donations to Public Goods: A Theory of Warm-Glow Giving." *Economic Journal,* 100: 464.

Aronson, Elliot. 1995. *The Social Animal.* 7th edition. New York: W. H. Freeman.

Arrow, Kenneth. 1973. "The Theory of Discrimination." In Orley Ashenfelter and Albert Rees, eds., *Discrimination in Labor Markets.* Princeton, N.J.: Princeton University Press.

Augustine. 1961. *The Confessions.* New York: Penguin Books.

Austin, Elizabeth. 1999. "Marks of Mystery; Psychological Reaction to Scars." Psychology Today. July 1, p. 46.

Axelrod, Robert. 1984. *The Evolution of Cooperation.* New York: Basic Books.

Ayres, Ian. 1987. "How Cartels Punish: A Structural Theory of Self-Enforcing Collusion." *Columbia Law Review,* 87: 295.

——— 1991. "The Possibility of Inefficient Corporate Contracts." *University of Cincinnati Law Review,* 60: 387.

Ayres, Ian, and Robert Gertner. 1989. "Filling Gaps in Incomplete Contracts: An Economic Theory of Default Rules." *Yale Law Journal,* 99: 87.

Ayres, Ian, and Barry J. Nalebuff. 1997. "Common Knowledge as a Barrier to Negotiation." *UCLA Law Review,* 44: 1631.

Bagwell, Laurie Simon, and B. Douglas Bernheim. 1996. "Veblen Effects in a Theory of Conspicuous Consumption." *American Economic Review,* 86: 349.

Baird, Douglas G., Robert H. Gertner, and Randal C. Picker. 1994. *Game Theory and the Law.* Cambridge, Mass.: Harvard University Press.

Baird, Douglas G., and Thomas H. Jackson. 1985. "Fraudulent Conveyance Law and Its Proper Domain." *Vanderbilt Law Review,* 38: 829.

Baird, Douglas G., and Robert Weisberg. 1982. "Rules, Standards, and the Battle of the Forms: A Reassessment of § 2–207." *Virginia Law Review,* 68: 1217.

Banerjee, Abhijit V. 1992. "A Simple Model of Herd Behavior." *Quarterly Journal of Economics,* 107: 797.

Banks, Jeffrey. 1991. *Signaling Games in Political Science.* New York: Harwood Academic.

Bartlett, Robert. 1986. *Trial by Fire and Water: The Medieval Judicial Ordeal.* Oxford: Clarendon Press.

Beattie, J. M. 1986. *Crime and the Courts in England, 1660–1800.* Princeton, N.J.: Princeton University Press.

Becker, Gary S. 1968. "Crime and Punishment: An Economic Approach." *Journal of Political Economy,* 76: 169.

——— 1971. *The Economics of Discrimination.* 2nd edition. Chicago: University of Chicago Press.

——— 1991. *A Treatise on the Family.* Enlarged edition. Cambridge, Mass.: Harvard University Press.

——— 1996. *Accounting for Tastes.* Cambridge, Mass.: Harvard University Press.

Becker, Gary S., Michael Grossman, and Kevin M. Murphy. 1991. "Rational Addiction and the Effect of Price on Consumption." *AEA Papers and Proceedings,* 81: 237.

Becker, Gary S., and Kevin M. Murphy. 1988. "A Theory of Rational Addiction." *Journal of Political Economy,* 96: 675.

Bell, Daniel. 1996. *The Cultural Contradictions of Capitalism.* New York: Basic Books.

Benson, Bruce L. 1990. *The Enterprise of Law: Justice Without the State.* San Francisco, Calif.: Pacific Research Institute for Public Policy.

——— 1992. "Customary Law as a Social Contract: International Commercial Law." *Constitutional Political Economy,* 3: 1.

——— 1995. "An Exploration of the Impact of Modern Arbitration Statutes on the Development of Arbitration in the United States." *Journal of Law, Economics, and Organization,* 11: 479.

Bergstrom, Theodore, Lawrence Blume, and Hal Varian. 1986. "On the Private Provision of Public Goods." *Journal of Public Economics,* 29: 25.

Bernheim, B. Douglas. 1994. "A Theory of Conformity." *Journal of Political Economy,* 102: 841.

Bernheim, B. Douglas, and Oded Stark. 1988. "Altruism Within the Family Reconsidered: Do Nice Guys Finish Last?" *American Economic Review,* 78: 1034.

Bernstein, Lisa. 1992. "Opting Out of the Legal System: Extralegal Contractual Relations in the Diamond Industry." *Journal of Legal Studies,* 21: 115.

——— 1996. "Merchant Law in a Merchant Court: Rethinking the Code's Search for Immanent Business Norms." *University of Pennsylvania Law Review,* 144: 1765.

——— 1999. "Private Commercial Law in the Cotton Industry: Value Creation Through Norms, Laws, and Institutions." Unpublished manuscript. University of Chicago Law School.

Besley, Timothy, and Stephen Coate. 1992. "Understanding Welfare Stigma: Taxpayer Resentment and Statistical Discrimination." *Journal of Public Economics,* 48: 165.

Bikhchandani, Suhil, David Hirshleifer, and Ivo Welch. 1992. "A Theory of Fads, Fashion, Custom, and Cultural Change as Informational Cascades." *Journal of Political Economy,* 100: 992.

Bishop, William. 1984. "Is He Married? Marriage and Information." *University of Toronto Law Review,* 34: 245.

Bix, Brian. 1998. "Bargaining in the Shadow of Love: The Enforcement of Premarital Agreements and How We Think About Marriage." *William and Mary Law Review,* 40: 145.

Blau, Peter M. 1964. *Exchange and Power in Social Life.* New York: J. Wiley.

Bowles, Samuel. 1998. "Endogenous Preferences: The Cultural Consequences of Markets and Other Economic Institutions." *Journal of Economic Literature,* 36: 75.

Boyd, Robert, and Peter J. Richerson. 1985. *Culture and the Evolutionary Process.* Chicago: University of Chicago Press.

Braithwaite, John. 1989. *Crime, Shame, and Reintegration.* Cambridge: Cambridge University Press.

Braithwaite, Valerie, and Margaret Levi, eds. 1998. *Trust and Governance.* New York: Russell Sage Foundation.

Brody, R. A., and B. I. Page. 1973. "Indifference, Alienation and Rational Decisions: The Effect of Candidate Evaluations on Turnout and Vote." *Public Choice,* 15: 1.

Buchanan, James M. 1968. *The Demand and Supply of Public Goods.* Chicago: Rand McNally.

——— 1975. "The Samaritan's Dilemma." In Edmund S. Phelps, ed., *Altruism, Morality, and Economic Theory.* New York: Russell Sage Foundation.

——— 1977. *Freedom in Constitutional Contract: Perspectives of a Political Economist.* College Station: Texas A&M University Press.

Burnham, John C. 1993. *Bad Habits: Drinking, Smoking, Taking Drugs, Gambling, Sexual Misbehavior, and Swearing in American History.* New York: New York University Press.

Bushway, Shawn D. 1997. "Labor-Market Effects of Permitting Employer Access to Criminal History Records." Unpublished manuscript, University of Maryland.

Camerer, Colin. 1988. "Gifts as Economic Signals and Social Symbols." *American Journal of Sociology,* 94: S180.

Carr, Jack L. and Janet T. Landa. 1983. "The Economics of Symbols, Clan Names, and Religion." *Journal of Legal Studies,* 12: 135.

Chaloupka, Frank. 1991. "Rational Addictive Behavior and Smoking." *Journal of Political Economy,* 99: 722.

Chang, Ruth 1997. "Introduction." In Ruth Chang, ed., *Incommensurability, Incomparability, and Practical Reason.* Cambridge, Mass.: Harvard University Press.

Cheung, Steven N. S. 1973. "The Fable of the Bees: An Economic Investigation." *Journal of Law and Economics,* 16: 11.

Clotfelter, Charles, ed., 1992. *Who Benefits from the Nonprofit Sector?* Chicago: University of Chicago Press.

Coate, Stephen, and Glenn C. Loury. 1993. "Will Affirmative-Action Policies Eliminate Negative Stereotypes?" *American Economic Review,* 83: 1220.

Coleman, James. 1990. *Foundations of Social Theory.* Cambridge, Mass.: Belknap Press of Harvard University Press.

Congleton, Roger D. 1989. "Efficient Status Seeking: Externalities, and the Evolution of Status Games." *Journal of Economic Behavior and Organization,* 11: 175.

Conlisk, John. 1996. "Why Bounded Rationality?" *Journal of Economic Literature,* 34: 669.

Conway, David. 1996. "Capitalism and Community." In Paul, Ellen Frankel, Fred D. Miller, Jr., and Jeffrey Paul, eds., *The Communitarian Challenge To Liberalism.* Cambridge: Cambridge University Press.

Coomer, Ken. 1970. "Three Recurrent and Acute Problems in Franchising." In C. Vaughn, ed., *Franchising Today.* Lynbrook, N.Y.: Farnsworth Publishing Company, Inc.

Cooter, Robert. 1994a. "Market Affirmative Action." *San Diego Law Review,* 31: 133.

——— 1994b. "Structural Adjudication and the New Law Merchant: A Model of Decentralized Law." *International Review of Law and Economics,* 14: 215.

——— 1996. "Decentralized Law for a Complex Economy: The Structural Approach to Adjudicating the New Law Merchant." *University of Pennsylvania Law Review,* 144: 1644.

——— 1998a. "Normative Failure Theory of the Law." *Cornell Law Review,* 82: 947.

——— 1998b. "Expressive Law and Economics." *Journal of Legal Studies,* 27: 585.

Cooter, Robert, and Janet T. Landa. 1984. "Personal versus Impersonal Trade: The Size of Trading Groups and Contract Law." *International Review of Law and Economics,* 4: 15.

Cox, Donald. 1987. "Motives for Private Income Transfers." *Journal of Political Economy,* 95: 508.

Craswell, Richard. 1988. "Contract Remedies, Renegotiation, and the Theory of Efficient Breach." *Southern California Law Review,* 61: 629.

——— 1998a. "Incommensurability, Welfare Economics, and the Law." *University of Pennsylvania Law Review,* 146: 1371.

——— 1999. "Do Trade Customs Exist?" In Jody Kraus and Steven Walt, eds., *The Jurisprudence of Corporate and Commercial Law.* Cambridge: Cambridge University Press (forthcoming).

Craswell, Richard, and John E. Calfee. 1986. "Deterrence and Uncertain Legal Standards." *Journal of Law, Economics, and Organization,* 2: 279.

Crotty, William. 1991. "Political Participation: Mapping the Terrain." In William Crotty, ed., *Political Participation and American Democracy.* New York: Greenwood Press.

Crowther, M. A. 1981. *The Workhouse System 1834–1929: The History of an English Social Institution.* Athens, Ga.: University of Georgia Press.

Davis, Natalie Zemon. 1975. *Society and Culture in Early Modern France: Eight Essays.* Stanford, Calif.: Stanford University Press.

"Dead Woman Forces Runoff In Senate Race." 1998. *The New York Times,* Aug. 27, p. A14.

D'Emilio, John. 1983. *Sexual Politics, Sexual Communities: The Making of a Homosexual Minority in the United States, 1940–1970.* Chicago: University of Chicago Press.

Demsetz, Harold. 1967. "Toward a Theory of Property Rights." *American Economic Review,* 57: 347.

——— 1988. *Ownership, Control, and the Firm.* Oxford: Basil Blackwell.

Diehm, John L. 1992. "Federal Expungement: A Concept in Need of a Definition." *Saint John's Law Review,* 66: 73.

Djilas, Aleksa. 1991. *The Contested Country: Yugoslav Unity and Communist Revolution, 1919–1953.* Cambridge, Mass.: Harvard University Press.

Douglas, Mary, and Aaron Wildavsky. 1982. *Risk and Culture: An Essay on the Selection of Technical and Environmental Dangers.* Berkeley: University of California Press.

Durden, Garey C., and Patricia Gaynor. 1987. "The Rational Behavior Theory of Voting Participation: Evidence from the 1970 and 1982 Elections." *Public Choice,* 53: 231.

Dworkin, Gerald. 1988. *The Theory and Practice of Autonomy.* Cambridge: Cambridge University Press.

Dwyer, John P., and Peter S. Menell. 1998. *Property Law and Policy: A Comparative Institutional Perspective.* Westbury, N.Y.: Foundation Press.

Eisenberg, Melvin Aron. 1979. "Donative Promises." *University of Chicago Law Review,* 47: 1.

Elias, Norbert. 1982. *The Civilizing Process.* Trans. Edmund Jephcott. New York: Pantheon Books.

Ellickson, Robert C. 1991. *Order Without Law: How Neighbors Settle Disputes.* Cambridge, Mass.: Harvard University Press.

——— 1993. "Property in Land." *Yale Law Journal,* 102: 1315.

Ellison, Glenn. 1994. "Cooperation in the Prisoner's Dilemma with Anonymous Random Matching." *Review of Economic Studies,* 61: 567.

Elster, Jon. 1989. *The Cement of Society: The Study of Social Order.* Cambridge: Cambridge University Press.

——— ed. 1998a. *Deliberative Democracy.* Cambridge: Cambridge University Press.

——— 1998b. "Emotions and Economic Theory." *Journal of Economic Literature,* 36: 47.

Epstein, Richard A. 1992a. *Forbidden Grounds: The Case Against Employment Discrimination Laws.* Cambridge, Mass.: Harvard University Press.

——— 1992b. "*International News Service vs. Associated Press:* Custom and the Law as Sources of Property Rights in News." *Virginia Law Review,* 78: 85.

——— 1992c. "The Path To *The T. J. Hooper:* The Theory and History of Custom in the Law of Tort." *Journal of Legal Studies,* 21: 1.

Eskridge, William N., Jr. 1996. *The Case for Same-Sex Marriage: From Sexual License to Civilized Commitment.* New York: Free Press.

Farnsworth, E. Allan. 1995. "Promises to Make Gifts." *American Journal of Comparative Law,* 43: 359.

Farrell, Joseph, and Matthew Rabin. 1996. "Cheap Talk." *Journal of Economic Perspectives,* 10: 103.

Frank, Robert H. 1985. *Choosing the Right Pond: Human Behavior and the Quest for Status.* New York: Oxford University Press.

——— 1988. *Passions Within Reason: The Strategic Role of Emotions.* New York: Norton.

Frank, Robert H., and Philip J. Cook. 1995. *The Winner-Take-All Society: Why the Few at the Top Get So Much More Than the Rest of Us.* New York: Penguin Books.

Friedman, David. 1988. "Does Altruism Produce Efficient Outcomes? Marshall versus Kaldor." *Journal of Legal Studies,* 17: 1.

Fudenberg, Drew, and Jean Tirole. 1991. *Game Theory.* Cambridge, Mass.: M. I. T. Press.

Fullbrook, Mary. 1997. "Myth-Making and National Identity: The Case of the GDR." In Geoffrey Hosking and George Schöpflin, eds., *Myths and Nationhood.* New York: Routledge.

Fuller, Lon L. 1941. "Consideration and Form." *Columbia Law Review,* 41: 799.

Furnas, J. C. 1959. *The Road to Harpers Ferry.* New York: William Sloane Associates.

Gambetta, Diego, ed. 1988. *Trust: Making and Breaking Cooperative Relations.* Oxford: Blackwell.

——— 1998. "'Claro!': An Essay on Discursive Machismo." In Jon Elster, ed., *Deliberative Democracy.* Cambridge: Cambridge University Press.

Gatrell, V. A. C. 1994. *The Hanging Tree: Execution and the English People, 1770–1868.* Oxford: Oxford University Press.

Gellner, Ernest. 1987. *Culture, Identity, and Politics.* Cambridge: Cambridge University Press.

——— 1992. *Reason and Culture: The Historic Role of Rationality and Rationalism.* Oxford: Blackwell.

——— 1994a. *Conditions of Liberty: Civil Society and Its Rivals.* London: Hamish Hamilton.

——— 1994b. *Encounters with Nationalism.* Oxford: Blackwell.

Gibbons, Robert. 1992. *Game Theory for Applied Economists.* Princeton, N.J.: Princeton University Press.

Gillette, Clayton P. 1998. "Lock-In Effects in Law and Norms." *Boston University Law Review,* 78: 813.

Gillis, John R. 1996. *A World of Their Own Making: Myth, Ritual, and the Quest for Family Values.* New York: Basic Books.

Glaeser, Edward L. 1998. "Economic Approach to Crime and Punishment." In Peter Newman, ed., *The New Palgrave Dictionary of Economics and the Law.* London: Macmillan.

Glaeser, Edward L., Bruce Sacerdote, and Jose A. Scheinkmann. 1996. "Crime and Social Interactions." *Quarterly Journal of Economics,* 111: 507.

Glazer, Amihai and Kai A. Konrad. 1996. "A Signaling Explanation for Charity." *American Economic Review,* 86: 1019.

Goetz, Charles J., and Robert E. Scott. 1980. "Enforcing Promises: An Examination of the Basis of Contract." *Yale Law Journal,* 89: 1261.

——— 1981. "Principles of Relational Contracts." *Virginia Law Review,* 67: 1089.

——— 1985. "The Limits of Expanded Choice: An Analysis of the Interactions Between Express and Implied Contract Terms." *California Law Review,* 73: 261.

Goffman, Erving. 1963. *Stigma: Notes on the Management of Spoiled Identity.* Englewood Cliffs, N.J.: Prentice Hall.

——— 1959. *The Presentation of Self in Everyday Life.* Woodstock, N.Y.: The Overlook Press.

Goldstein, Robert Justin. 1996. *Burning the Flag: The Great 1989–1990 American Flag Desecration Controversy.* Kent, Ohio: Kent State University Press.

Gordley, James. 1991. *The Philosophical Origins of Modern Contract Doctrine.* Oxford: Clarendon Press.

Green, Donald P., and Ian Shapiro. 1994. *Pathologies of Rational Choice Theory: A Critique of Applications in Political Science.* New Haven: Yale University Press.

Greenberg, Kenneth S. 1996. *Honor and Slavery.* Princeton, N.J.: Princeton University Press.

Greif, Avner. 1993. "Contract Enforceability and Economic Institutions in Early Trade: The Maghribi Traders Coalition." *American Economic Review,* 83: 525.

Greif, Avner, Paul Milgrom, and Barry R. Weingast. 1994. "Coordination, Commitment, and Enforcement: The Case of the Merchant Guild." *Journal of Political Economy,* 102: 745.

Gustafson, W. Mark. 1997. "Inscripta in fronte: Penal Tattooing in Late Antiquity." *Classical Antiquity,* 16: 79.

Habermas, Jürgen. 1984. *The Theory of Communicative Action.* Volume 1. Trans. Thomas McCarthy. Boston: Beacon Press.

Hackett, Steven C. 1994. "Is Relational Exchange Possible in the Absence of Reputations and Repeated Contact?" *Journal of Law, Economics, and Organization,* 10: 360.

Hadfield, Gillian. 1990. "Problematic Relations: Franchising and the Law of Incomplete Contracts." *Stanford Law Review,* 42: 927.

——— 1992. "Bias in the Evolution of Legal Rules." *Georgetown Law Journal,* 80: 583.

——— 1994. "Judicial Competence and the Interpretation of Incomplete Contracts." *Journal of Legal Studies,* 23: 159.

Hansmann, Henry. 1996. *The Ownership of Enterprise.* Cambridge, Mass.: The Belknap Press of Harvard University Press.

Hardin, Russell. 1982. *Collective Action.* Baltimore: Johns Hopkins University Press.

——— 1995. *One for All: The Logic of Group Conflict.* Princeton, N.J.: Princeton University Press.

Harrington, Joel. 1995. *Reordering Marriage and Society in Reformation Germany.* Cambridge: Cambridge University Press.

Hart, H. L. A. 1983. *Essays in Jurisprudence and Philosophy.* Oxford: Clarendon Press.

Hasen, Richard L. 1996. "Voting Without Law." *University of Pennsylvania Law Review,* 144: 2135.

Hawthorne, Nathaniel. 1981. *The Scarlet Letter.* Toronto: Bantam Books.

Hayek, F. H. 1973. *Law, Legislation, and Liberty: A New Statement of the Liberal Principles of Justice and Political Economy.* Volume 3. Chicago: University of Chicago Press.

Hebdige, Dick. 1979. *Subculture, The Meaning of Style.* London: Methuen.

Hechter, Michael. 1987. *Principles of Group Solidarity.* Berkeley: University of California Press.

Hedges, Chris. 1998. "Dejected Belgrade Embraces Hedonism, but Still, Life Is No Cabaret." *New York Times,* Jan. 19, p. A1.

Helmholz, R. H. 1974. *Marriage Litigation in Medieval England.* Cambridge: Cambridge University Press.

Higgins, Richard A., and Paul H. Rubin. 1986. "Counterfeit Goods." *Journal of Law and Economics,* 29: 211.

Hill, Thomas E. 1991. *Autonomy and Self-Respect.* Cambridge: Cambridge University Press.

Hirsch, Fred 1976. *Social Limits to Growth.* Cambridge, Mass.: Harvard University Press.

Hirschman, Albert O. 1982. *Shifting Involvements: Private Interest and Public Action.* Princeton: Princeton University Press.

——— 1986. *Rival Views of Market Society.* Cambridge, Mass.: Harvard University Press.

Hirshleifer, David, and Eric Rasmusen. 1989. "Cooperation in a Repeated Prisoners' Dilemma with Ostracism." *Journal of Economic Behavior and Organization,* 12: 87.

Hirshleifer, Jack. 1982. "Evolutionary Models in Economics and the Law: Cooperation versus Conflict Strategies." *Research in Law and Economics,* 4: 1.

——— 1987a. *Economic Behaviour in Adversity.* Chicago: University of Chicago Press.

——— 1987b. "On Emotions as Guarantors of Threats and Promises." In John Dupre, ed., *The Latest on the Best: Essays on Evolution and Optimality.* Cambridge, Mass.: M. I. T. Press.

Hobsbawm, Eric J. 1990. *Nations and Nationalism Since 1780: Programme, Myth, Reality.* Cambridge: Cambridge University Press.

Hobsbawm, Eric J., and Terence Ranger, eds. 1983. *The Invention of Tradition.* Cambridge: Cambridge University Press.

Holmes, Oliver Wendell. 1963. *The Common Law.* M. Howe, ed. Cambridge, Mass.: The Belknap Press of Harvard University Press.

Holmes, Stephen. 1995. *Passions and Constraint: On the Theory of Liberal Democracy.* Chicago: University of Chicago Press.

Huang, Peter and Ho Mou Wu. 1994. "More Order Without Law: A Theory of Social Norms and Organizational Cultures." *Journal of Law, Economics, and Organization,* 10: 390.

Iannaccone, Lawrence R. 1992. "Sacrifice and Stigma: Reducing Free-Riding in Cults, Communes, and Other Collectives." *Journal of Political Economy,* 100: 271.

Ingram, Martin. 1984. "Ridings, Rough Music and the 'Reform of Popular Culture' in Early Modern England." *Past and Present,* 105: 79.

Jackson, Robert A. 1995. "Clarifying the Relationship Between Education and Turnout." *American Politics Quarterly,* 23: 279.

Jankowski, Martin Sanchez. 1991. *Islands in the Street: Gangs and American Urban Society.* Berkeley: University of California Press.

Johnsen, D. Bruce. 1986. "The Formation and Protection of Property Rights Among the Southern Kwakiutl Indians." *Journal of Legal Studies,* 15: 41.

Johnston, Jason Scott. 1991. "Uncertainty, Chaos, and the Torts Process: An Economic Analysis of Legal Form." *Cornell Law Review,* 76: 341.

Jones, C. P. 1987. "Stigma: Tattooing and Branding in Graeco-Roman Antiquity." *Journal of Roman Studies,* 78: 139.

Jones, Stephen R. G. 1984. *The Economics of Conformism.* Oxford: Blackwell.

Kagan, Robert A., and Jerome H. Skolnick. 1993. "Banning Smoking: Compliance Without Enforcement." In Robert L. Rabin and Stephen D. Sugarman, eds., *Smoking Policy: Law, Politics, and Culture.* New York: Oxford University Press.

Kagel, John H., and Alvin E. Roth, eds. 1995. *The Handbook of Experimental Economics.* Princeton, N.J.: Princeton University Press.

Kahan, Dan M. 1996. "What Do Alternative Sanctions Mean?" *University of Chicago Law Review,* 63: 591.

——— 1997. "Social Influence, Social Meaning, and Deterrence." *University of Virginia Law Review,* 83: 349.

Kahan, Dan M., and Eric A. Posner. 1999. "Shaming White Collar Criminals: A Proposal for Reform of the Federal Sentencing Guidelines." *Journal of Law and Economics,* 42: 365.

Kan, Sergei. 1989. *Symbolic Immortality: The Tlingit Potlatch of the Nineteenth Century.* Washington: Smithsonian Institution Press.

Kandori, Michihiro. 1992. "Social Norms and Community Enforcement." *Review of Economic Studies,* 59: 63.

Kaplow, Louis. 1992. "Rules Versus Standards: An Economic Analysis." *Duke Law Journal,* 42: 557.

——— 1995. "A Note on Subsidizing Gifts." *Journal of Public Economics,* 58: 469.

Kaplow, Louis, and Steven Shavell. 1994. "Why the Legal System Is Less Efficient Than the Income Tax in Redistributing Income." *Journal of Legal Studies,* 23: 667.

Karpoff, Jonathan M. and John R. Lott, Jr. 1993. "The Reputational Penalty Firms Bear from Committing Criminal Fraud." *Journal of Law and Economics,* 36: 757.

Katz, Leo. 1996. *Ill-Gotten Gains: Evasion, Fraud, Blackmail, and Kindred Puzzles of the Law.* Chicago: University of Chicago Press.

Kelman, Steven. 1981. "Cost-Benefit Analysis: An Ethical Critique." *Regulation,* 5: 33.

Kennedy, Duncan. 1973. "Legal Formality." *Journal of Legal Studies,* 2: 351.

Kiernan, V. G. 1988. *The Duel in European History: Honor and the Reign of Aristocracy.* Oxford: Oxford University Press.

Klein, Benjamin, and Keith B. Leffler. 1981. "The Role of Market Forces in Assuring Contractual Performance." *Journal of Political Economy,* 89: 615.

Klein, Richard. 1997. "After the Preaching, the Lure of the Taboo." *The New York Times,* Aug. 24, section 2, p. 1.

Klier, John D. 1997. "The Myth of Zion among East European Jewry." In Geoffrey Hosking and George Schöpflin, eds., *Myths and Nationhood.* New York : Routledge.

Knack, Stephen. 1992. "Civic Norms, Social Sanctions and Voter Turnout." *Rationality and Society,* 4: 133.

Kniss, Fred LaMar. 1997. *Disquiet in the Land: Cultural Conflict in American Mennonite Communities.* New Brunswick, N.J.: Rutgers University Press.

Koford, Kenneth J., and Jeffrey B. Miller, eds. 1991. *Social Norms and Economic Institutions.* Ann Arbor: University of Michigan Press.

Koppelman, Andrew. 1996. *Antidiscrimination Law and Social Equality.* New Haven, Conn.: Yale University Press.

Kranton, Rachel E. 1996a. "The Formation of Cooperative Relationships." *Journal of Law, Economics, and Organization,* 12: 214.

——— 1996b. "Reciprocal Exchange: A Self-Sustaining System." *American Economic Review,* 86: 830.

Kraus, Jody S. 1997. "Legal Design and the Evolution of Commercial Norms." *Journal of Legal Studies,* 26: 337.

Kreps, David. 1990a. "Corporate Culture and Economic Theory." In James E. Alt and Kenneth A. Shepsle, eds., *Perspectives on Positive Political Economy.* Cambridge: Cambridge University Press.

——— 1990b. *A Course in Microeconomic Theory.* Princeton, N.J.: Princeton University Press.

——— 1990c. *Game Theory and Economic Modelling.* Oxford: Clarendon Press.

Kreps, David, Paul Milgrom, John Roberts, and Robert Wilson. 1982. "Rational Cooperation in the Finitely Repeated Prisoners' Dilemma." *Journal of Economic Theory,* 27: 245.

Kronman, Anthony. 1985. "Contract Law and the State of Nature." *Journal of Law, Economics, and Organization,* 1: 5.

Kull, Andrew. 1992. "Reconsidering Gratuitous Promises." *Journal of Legal Studies,* 21: 39.

Kuran, Timur. 1995. *Private Truths, Public Lies.* Cambridge, Mass.: Harvard University Press.

——— 1998. "Ethnic Norms and Their Transformation through Reputational Cascades." *Journal of Legal Studies,* 27: 623.

Kuran, Timur, and Cass R. Sunstein. 1999. "Availability Cascades and Risk Regulation." *Stanford Law Review,* 51: 683.

Kurtz, Howard. 1998. "A Tough Sell for David Brinkley; Colleagues Are Uneasy With Ex-Newsman's Enterprising Role." *Washington Post,* Jan. 8, p. B1.

Kymlicka, Will. 1996. "Social Unity in a Liberal State." In Paul, Ellen Frankel, Fred D. Miller, Jr., and Jeffrey Paul, eds., *The Communitarian Challenge To Liberalism.* Cambridge: Cambridge University Press.

Landa, Janet T. 1994. *Trust, Ethnicity, and Identity: Beyond the New Institutional Economics of Ethnic Trading Networks, Contract Law, and Gift Exchange.* Ann Arbor: University of Michigan Press.

Landes, William M., and Richard A. Posner. 1987. "Trademark Law: An Economic Perspective." *Journal of Law and Economics,* 30: 265.

Lea, Henry Charles. 1967. "The Wager of Battle." In Paul Bohannan, ed., *Law and Warfare: Studies in the Anthropology of Conflict.* Garden City, N.Y.: The Natural History Press.

Leacock, Eleanor. 1954. "The Montagnais 'Hunting Territory' and the Fur Trade." *American Anthropologist,* 56: Memoir No. 78, 1.

Leighley, Jan E., and Jonathan Nagler. 1992. "Individual and Systematic Influences on Turnout: Who Votes? 1984." *Journal of Politics,* 54: 718.

Lessig, Lawrence 1995. "The Regulation of Social Meaning." *University of Chicago Law Review,* 62: 943.

Levine, David, and Keith Wrightson. 1980. "The Social Control of Illegitimacy in Early Modern England." In Peter Laslett, Karla Costerveen, and Richard M. Smith, eds., *Bastardy and Its Comparative History: Studies in the History of Illegitimacy and Marital Nonconformism in Britain, France, Germany, Sweden, North America, Jamaica, and Japan.* Cambridge, Mass.: Harvard University Press.

Levmore, Saul. 1995. "Love It or Leave It: Property Rules, Liability Rules, and Exclusivity of Remedies in Partnership and Marriage." *Law and Contemporary Problems,* 58: 221.

Lewis, David K. 1969. *Convention: A Philosophical Study.* Cambridge, Mass.: Harvard University Press.

Lewis, W. H. 1957. *The Splendid Century: Life in the France of Louis XIV.* Garden City, N.Y.: Doubleday Anchor.

Lindbeck, Assar, Sten Nyberg, and Jörgen W. Weibull. 1999. "Social Norms and Economic Incentives in the Welfare State." *Quarterly Journal of Economics,* 64: 1.

Llewellyn, Karl. 1931. "What Price Contract?—An Essay in Perspective." *Yale Law Journal,* 40: 704.

Lott, John R., Jr. 1992. "Do We Punish High-Income Criminals Too Heavily?" *Economic Inquiry*, 30: 583.
Loury, Glenn C. 1994. "Self-Censorship in Public Discourse." *Rationality and Society*, 6: 428.
Lundberg, Shelly, and Robert Pollak. 1993. "Separate Spheres Bargaining and the Marriage Market." *Journal of Political Economy*, 101: 988.
Lupia, Arthur, and Mathew D. McCubbins. 1998. *The Democratic Dilemma: Can Citizens Learn What They Need to Know?* Cambridge: Cambridge University Press.
Macaulay, Stewart. 1963. "Non-Contractual Relations in Business: A Preliminary Study." *American Sociological Review*, 28: 55.
Macedo, Stephen. 1996. "Community, Diversity, and Civic Education: Toward a Liberal Political Science of Group Life." In Paul, Ellen Frankel, Fred D. Miller, Jr., and Jeffrey Paul, eds., *The Communitarian Challenge To Liberalism.* Cambridge: Cambridge University Press.
Macneil, Ian R. 1978. "Contracts: Adjustment of Long-Term Economic Relations Under Classical, Neoclassical, and Relational Contract Law." *Northwestern University Law Review*, 72: 854.
Madow, Michael. 1993. "Private Ownership of Public Image: Popular Culture and Publicity." *California Law Review*, 81: 127.
Mailath, George J. 1998. "Do People Play Nash Equilibrium? Lessons from Evolutionary Game Theory." *Journal of Economic Literature*, 36: 1347.
Malcolm, Noel. 1996. *Bosnia: A Short History.* New York: New York University Press.
Margolis, Howard. 1982. *Selfishness, Altruism, and Rationality: A Theory of Social Choice.* Cambridge: Cambridge University Press.
Mauss, Marcel. 1990. *The Gift: The Form and Reason for Exchange in Archaic Societies.* Trans. W. D. Halls. London: Routledge.
Maynard Smith, John. 1982. *Evolution and the Theory of Games.* Cambridge: Cambridge University Press.
McAdams, Richard H. 1992. "Relative Preferences." *Yale Law Journal*, 102: 1.
——— 1995. "Cooperation and Conflict: The Economics of Group Status Production and Race Discrimination." *Harvard Law Review*, 108: 1003.
——— 1996. "Group Norms, Gossip, and Blackmail." *University of Pennsylvania Law Review*, 144: 2237.
——— 1997. "The Origin, Development, and Regulation of Norms." *Michigan Law Review*, 96: 338.
Meredith, Robyn. 1997. "Strip Clubs Under Siege as Salesman's Havens." *New York Times*, Sept. 20, p. A1.
Milgrom, Paul, Douglass C. North, and Barry R. Weingast. 1990. "The Role of Institutions in the Revival of Trade: The Medieval Law Merchant, Private Judges, and the Champagne Fairs." *Economics and Politics*, 2: 1.
Milgrom, Paul and John Roberts. 1986. "Price and Advertising Signals of Product Quality." *Journal of Political Economy*, 94: 796.
Miller, William Ian. 1990. *Bloodtaking and Peacemaking: Feud, Law, and Society in Saga Iceland.* Chicago: University of Chicago Press.
——— 1993. *Humiliation: And Other Essays on Honor, Social Discomfort, and Violence.* Ithaca, N.Y.: Cornell University Press.
Moffitt, Robert. 1983. "An Economic Model of Welfare Stigma." *American Economic Review*, 73: 1023.

Morris, Stephen. 1998. *An Instrumental Theory of Political Correctness.* Unpublished manuscript. University of Pennsylvania.
Mosse, George L. 1978. *Toward the Final Solution: A History of European Racism.* New York: H. Fertig.
——— 1985. *Nationalism and Sexuality: Respectability and Abnormal Sexuality in Modern Europe.* New York: Howard Fertig.
Mueller, Dennis C. 1989. *Public Choice II.* Cambridge: Cambridge University Press.
Murphy, Richard S. 1996. "Property Rights in Personal Information: An Economic Defense of Privacy." *Georgetown Law Journal,* 84: 2381.
Neilson, George. 1890. *Trial by Combat.* Glasgow: William Hodge & Co.
Netanyahu, B. 1995. *The Origins of the Inquisition in Fifteenth Century Spain.* New York: Random House.
Nissenbaum, Stephen. 1997. *The Battle for Christmas.* New York: Knopf.
North, Douglass C. 1990. *Institutions, Institutional Change, and Economic Performance.* Cambridge: Cambridge University Press.
Nussbaum, Martha C. 1997. "Flawed Foundations: The Philosophical Critique of (a Particular Type of) Economics." *University of Chicago Law Review,* 64: 1197.
——— 1999. *Women and Human Development: The Capabilities Approach* (forthcoming).
Olson, Mancur. 1965. *The Logic of Collective Action: Public Goods and the Theory of Groups.* Cambridge, Mass.: Harvard University Press.
Ostrom, Elinor. 1990. *Governing the Commons: The Evolution of Institutions for Collective Action.* Cambridge: Cambridge University Press.
Ostrower, Francie. 1995. *Why the Wealthy Give: The Culture of Elite Philanthropy.* Princeton, N.J.: Princeton University Press.
Paul, Ellen Frankel, Fred D. Miller, Jr., and Jeffrey Paul, eds. 1996. *The Communitarian Challenge To Liberalism.* Cambridge: Cambridge University Press.
Pesendorfer, Wolfgang. 1995. "Design Innovation and Fashion Cycles." *American Economic Review,* 85: 771.
Picker, Randal C. 1997. "Simple Games in a Complex World: A Generative Approach to the Adoption of Norms." *University of Chicago Law Review,* 64: 1225.
Piore, Michael J. 1995. *Beyond Individualism.* Cambridge, Mass.: Harvard University Press.
Posner, Eric A. 1995. "Contract Law in the Welfare State: A Defense of the Unconscionability Doctrine, Usury Laws, and Related Limitations on the Freedom to Contract." *Journal of Legal Studies,* 24: 283.
——— 1996a. "Law, Economics, and Inefficient Norms." *University of Pennsylvania Law Review,* 144: 1697.
——— 1996b. "The Legal Regulation of Religious Groups." *Legal Theory,* 2: 33.
——— 1996c. "The Regulation of Groups: The Influence of Legal and Nonlegal Sanctions on Collective Action." *University of Chicago Law Review,* 63: 133.
——— 1997a. "Altruism, Status, and Trust in the Law of Gifts and Gratuitous Promises." *Wisconsin Law Review,* 1997: 567.
——— 1997b. "Standards, Rules, and Social Norms." *Harvard Journal of Law and Public Policy,* 21: 101.
——— 1998a. "Efficient Norms." In Peter Newman, ed., *The New Palgrave Dictionary of Economics and the Law.* London: Macmillan Reference.
——— 1998b. "Symbols, Signals, and Social Norms in Politics and the Law." *Journal of Legal Studies,* 27: 765.

Posner, Richard A. 1977. "Gratuitous Promises in Economics and Law." *Journal of Legal Studies,* 6: 411.

——— 1981. *The Economics of Justice.* Cambridge, Mass.: Harvard University Press.

——— 1997. "Social Norms and the Law: An Economic Approach." *American Economic Review,* 87: 365.

——— 1998a. *Economic Analysis of Law.* 5th edition. New York: Aspen Law & Business.

——— 1998b. "Privacy." In Peter Newman, ed., *The New Palgrave Dictionary of Economics and the Law.* London: Macmillan Reference.

Post, Elizabeth L. 1992. 15th Edition. *Emily Post's Etiquette.* New York: HarperCollins Publishers.

Presser, Stanley, and Michael Traugott. 1992. "Little White Lies and Social Science Models." *Public Opinion Quarterly,* 56: 77.

Przeworski, Adam. 1998. "Deliberation and Ideological Domination." In Jon Elster, ed., *Deliberative Democracy.* Cambridge: Cambridge University Press.

Putnam, Robert. 1993. *Making Democracy Work: Civic Tradition in Modern Italy.* Princeton, N.J.: Princeton University Press.

Quintanilla, Carl. 1998. "Making a Gift Last? Better Think Twice About One for the Boss." *The Wall Street Journal,* Nov. 18, p. A1.

Rabin, Matthew. 1998. "Psychology and Economics." *Journal of Economic Literature,* 36: 11.

Radin, Margaret Jane. 1987. "Market-Inalienability." *Harvard Law Review,* 100: 1849.

Ramseyer, Mark. 1996. "Products Liability Through Private Ordering: Notes on a Japanese Experiment." *University of Pennsylvania Law Review,* 144: 1823.

Rasmusen, Eric. 1994. 2nd Edition. *Games and Information: An Introduction to Game Theory.* New York: Blackwell.

——— 1996. "Stigma and Self-Fulfilling Expectations of Criminality." *Journal of Law and Economics,* 39: 519.

——— 1998. "The Economics of Desecration: Flag Burning and Related Activities." *Journal of Legal Studies,* 27: 245.

Rawls, John. 1993. *Political Liberalism.* New York: Columbia University Press.

Raz, Joseph. 1986. *The Morality of Freedom.* Oxford: Clarendon Press.

——— 1990. *Practical Reason and Norms.* Princeton, N.J.: Princeton University Press.

——— 1994. *Ethics in the Public Domain: Essays in the Morality of Law and Politics.* Oxford: Clarendon Press.

Regan, Donald H. 1989. "Authority and Value: Reflections on Raz's Morality of Freedom." *Southern California Law Review,* 62: 995.

Roback, Jennifer. 1989. "Racism As Rent Seeking." *Economic Inquiry,* 27: 661.

Robinson, Thomas N., and Jodel D. Killen. 1997. "Do Cigarette Warning Labels Reduce Smoking?: Paradoxical Effects Among Adolescents." *Archives of Pediatric and Adolescent Medicine,* 151: 267.

Rock, Edward B. 1997. "Saints and Sinners: How Does Delaware Corporate Law Work?" *U. C. L. A. Law Review,* 44: 1009.

Romano, Roberta. 1987. "The Political Economy of Takeover Statutes." *Virginia Law Review,* 73: 111.

Roulet, Marguerite. 1996. "Dowry and Prestige in Northern India." *Contributions to Indian Sociology,* 30: 89.

Russell, Francis. 1986. *Sacco & Vanzetti: The Case Resolved.* New York: Harper & Row.

Sandel, Michael J. 1982. *Liberalism and the Limits of Justice.* Cambridge: Cambridge University Press.

Schelling, Thomas. 1960. *The Strategy of Conflict.* Cambridge, Mass.: Harvard University Press.

——— 1978. *Micromotives and Macrobehavior.* New York: Norton.

——— 1984. *Choice and Consequence.* Cambridge, Mass.: Harvard University Press.

Schmitt, Richard B. 1995. "Uncharitable Acts: If Donors Fail to Give, More Nonprofit Groups Take Them to Court." *Wall Street Journal,* July 27, at A1.

Schneewind, J. B. 1998. *The Invention of Autonomy: A History of Modern Moral Philosophy.* Cambridge: Cambridge University Press.

Schotter, Andrew. 1981. *The Economic Theory of Social Institutions.* Cambridge: Cambridge University Press.

Schumpeter, Joseph A. 1950. *Capitalism, Socialism and Democracy.* 3rd edition. New York: Harper.

Schwab, Stewart. 1986. "Is Statistical Discrimination Efficient?" *American Economic Review,* 76: 228.

Schwartz, Alan. 1990. "The Myth that Promisees Prefer Supracompensatory Remedies: An Analysis of Contracting for Damage Measures." *Yale Law Journal,* 100: 369.

——— 1992. "Relational Contracts in the Courts: An Analysis of Incomplete Agreements and Judicial Strategies." *Journal of Legal Studies,* 21: 271.

——— 1998. "Incomplete Contracts." In Peter Newman, ed., *The New Palgrave Dictionary of Economics and the Law.* London: Macmillan Reference.

Schwartz, Warren F., et al. 1984. "The Duel: Can These Gentlemen Be Acting Efficiently?" *Journal of Legal Studies,* 13: 321.

Scitovsky, Tibor. 1992. *The Joyless Economy: The Psychology of Human Satisfaction.* Revised Edition. New York: Oxford University Press.

Scott, Elizabeth S., and Robert E. Scott. 1998. "Marriage as Relational Contract." *Virginia Law Review,* 84: 1225.

Sen, Amartya. 1977. "Rational Fools: A Critique of the Behavioral Foundations of Economic Theory." *Philosophy and Public Affairs,* 6: 317.

Sethi, Rajiv, and E. Somanathan. 1996. "The Evolution of Social Norms in Common Property Resource Use." *American Economic Review,* 86: 766.

Shavell, Steven. 1991. "An Economic Analysis of Altruism and Deferred Gifts." *Journal of Legal Studies,* 20: 401.

Sherman, Lawrence W. 1993. "Defiance, Deterrence, and Irrelevance: A Theory of the Criminal Sanction." *Journal of Research in Crime and Delinquency,* 30: 445.

Silbaugh, Katharine B. 1998. "Marriage Contracts and the Family Economy." *Northwestern University Law Review,* 93: 65.

Simeone, William E. 1995. *Rifles, Blankets, and Beads: Identity, History, and the Northern Athapaskan Potlatch.* Norman, Oklahoma: University of Oklahoma Press.

Simmel, Georg. 1978. *The Philosophy of Money.* Trans. Tom Bottomore and David Frisby. London: Routledge and Kegan Paul.

Smith, Anthony. 1997. "The 'Golden Age' and National Renewal." In Geoffrey Hosking and George Schöpflin, eds., *Myths and Nationhood.* New York: Routledge.

Smith, Craig R. 1998. "Moon Cakes: Gifts That Keep on Giving and Giving and . . ." *Wall Street Journal,* Sept. 30, p. 1.

Sober, Elliot, and David Sloan Wilson. 1998. *Unto Others: The Evolution and Psychology of Unselfish Behavior.* Cambridge, Mass.: Harvard University Press.

Solow, John. 1993. "Is It Really the Thought That Counts: Toward a Rational Theory of Christmas." *Rationality and Society,* 5: 506.

Spence, A. Michael. 1974. *Market Signaling: Informational Transfer in Hiring and Related Screening Processes.* Cambridge, Mass.: Harvard University Press.

Spicker, Paul. 1984. *Stigma and Social Welfare.* London: Croom Helm.

Spierenburg, Pieter. 1995. "The Body and the State: Early Modern Europe." In Norval Morris and David J. Rothman, eds., *The Oxford History of the Prison.* Oxford: Oxford University Press.

Stark, Oded. 1995. *Altruism and Beyond: An Economic Analysis of Transfers and Exchanges within Families and Groups.* Cambridge: Cambridge University Press.

Sugden, Robert. 1982. "On the Economics of Philanthropy." *Economic Journal,* 92: 341.

——— 1986. *The Economics of Rights, Co-operation, and Welfare.* Oxford: Blackwell.

Sunstein, Cass. 1994. "Incommensurability and Valuation in Law." *Michigan Law Review,* 92: 779.

——— 1996. "Social Norms and Social Roles." *Columbia Law Review,* 96: 903.

Taylor, Charles. 1995. *Philosophical Arguments.* Cambridge, Mass.: Harvard University Press.

Taylor, Michael. 1976. *Anarchy and Cooperation.* London: Wiley.

——— 1982. *Community, Anarchy, and Liberty.* Cambridge: Cambridge University Press.

——— 1987. *The Possibility of Cooperation.* Cambridge: Cambridge University Press.

Teja, Mohinderjit Kaur. 1993. *Dowry: A Study in Attitudes and Practices.* New Delhi, India: Inter-India Publications.

Telser, Lester G. 1980. "A Theory of Self-Enforcing Agreements." *Journal of Business,* 53: 27.

Thaler, Richard H. 1991. *Quasi Rational Economics.* New York: Russell Sage Foundation.

Thompson, E. P. 1993. *Customs in Common.* New York: New York Press.

Titmuss, Richard M. 1971. *The Gift Relationship: From Human Blood to Social Policy.* New York: Vintage Books.

Trebilcock, Michael J. 1993. *The Limits of Freedom of Contract.* Cambridge, Mass.: Harvard University Press.

Trebilcock, Michael J., and Rosemin Keshvani. 1991. "The Role of Private Ordering in Family Law." *University of Toronto Law Journal,* 41: 533.

Trivers, Robert L. 1971. "The Evolution of Reciprocal Altruism." *Quarterly Review of Biology,* 46: 35.

Tyler, Tom R. 1990. *Why People Obey the Law.* New Haven, Conn.: Yale University Press.

Ullman-Margalit, Edna. 1977. *The Emergence of Norms.* Oxford: Clarendon Press.

Veblen, Thorstein. 1992. *The Theory of the Leisure Class.* New Brunswick, U.S.A.: Transaction Publishers.

Waldfogel, Joel. 1993. "The Deadweight Loss of Christmas." *American Economic Review,* 83: 1328.

——— 1994a. "Does Conviction Have a Persistent Effect on Income and Employment?" *International Review of Law and Economics,* 14: 103.

——— 1994b. "The Effect of Criminal Convictions on Income and the 'Trust Reposed in the Workman.'" *Journal of Human Resources,* 29: 62.

Waxman, Chaim Isaac. 1977. *The Stigma of Poverty: A Critique of Poverty Theories and Policies.* New York: Pergamon Press.

Weber, Eugene. 1976. *Peasants Into Frenchmen: The Modernization of Rural France, 1870–1914.* Stanford, Calif.: Stanford University Press.

West, Mark. 1997. "Legal Rules and Social Norms in Japan's Secret World of Sumo." *Journal of Legal Studies,* 26: 165.

Whitman, James Q. 1996. "The Moral Menace of Roman Law and the Making of Commerce: Some Dutch Evidence." *Yale Law Journal,* 105: 1841.

Wilentz, Sean, ed. 1985. *Rites of Power: Symbolism, Ritual, and Politics Since the Middle Ages.* Philadelphia: University of Pennsylvania Press.

Williamson, Oliver E. 1983. "Credible Commitments: Using Hostages to Support Exchange." *American Economic Review,* 73: 519.

——— 1985. *The Economic Institutions of Capitalism: Firms, Markets, Relational Contracting.* New York: Free Press.

Wu, Jianzhong, and Robert Axelrod. 1995. "How to Cope with Noise in the Iterated Prisoner's Dilemma." *Journal of Conflict Resolution* 39: 183.

Wyatt-Brown, Bertram. 1982. *Southern Honor: Ethics and Behavior in the Old South.* New York: Oxford University Press.

Young, H. Peyton. 1996. "The Economics of Convention." *Journal of Economic Perspectives,* 10: 105.

——— 1998a. *Individual Strategy and Social Structure: An Evolutionary Theory of Institutions.* Princeton, N.J.: Princeton University Press.

——— 1998b. "Social Norms and Economic Welfare." *European Economic Review,* 42: 821.

Zelizer, Viviana A. 1994. *The Social Meaning of Money: Pin Money, Paychecks, Poor Relief, and Other Currencies.* New York: Basic Books.

Acknowledgments

Thanks to Steve Choi, Richard Craswell, Emlyn Eisenach, Richard Epstein, Nuno Garoupa, Beth Garrett, Andrew Guzman, Dan Kahan, William Landes, John Lott, Tracey Meares, Geoffrey Miller, Martha Nussbaum, Randy Picker, Richard Posner, Ricky Revesz, Steve Schulhofer, David Strauss, Cass Sunstein, and participants at talks at the law schools of the University of Chicago, Stanford, Berkeley, New York University, the University of Michigan, the University of Minnesota, Georgetown, and Yale. Special thanks to Ian Ayres, Robert Ellickson, Jack Goldsmith, Douglas Lichtman, Richard McAdams, and Adrian Vermeule, who gave me comments on the entire manuscript; and again to Ellickson for his generosity, encouragement, and help over many years. Richard Izquierdo and Karen Schoen provided valuable research assistance.

Chapters 4, 5, 7, and 11 are heavily revised versions of "Altruism, Status, and Trust in the Law of Gifts and Gratuitous Promises," *Wisconsin Law Review* 1997: 567 (1997); "Family Law and Social Norms," in *The Fall and Rise of Freedom of Contract* (Frank Buckley, ed., Duke University Press, 1999); "Symbols, Signals, and Social Norms in Politics and the Law," *Journal of Legal Studies,* 27: 765 (1998); and "The Strategic Basis of Principled Behavior: A Critique of the Incommensurability Thesis," *University of Pennsylvania Law Review,* 146: 1185 (1998). In addition, fragments of "Law, Economics, and Inefficient Norms," *University of Pennsylvania Law Review,* 144: 1697 (1996), can be found in Chapter 10. My thanks to the copyright holders for permission to use this material.

Index